The Basic Physics of
Quantum
Theory

The Basic Physics of
Quantum Theory

Basil S Davis

Xavier University of Louisiana, USA

World Scientific

NEW JERSEY · LONDON · SINGAPORE · BEIJING · SHANGHAI · HONG KONG · TAIPEI · CHENNAI · TOKYO

Published by

World Scientific Publishing Co. Pte. Ltd.

5 Toh Tuck Link, Singapore 596224

USA office: 27 Warren Street, Suite 401-402, Hackensack, NJ 07601

UK office: 57 Shelton Street, Covent Garden, London WC2H 9HE

Library of Congress Cataloging-in-Publication Data
Names: Davis, Basil S. (Lecturer in physics), author.
Title: The basic physics of quantum theory / Basil S. Davis,
    Xavier University of Louisiana, USA.
Description: New Jersey : World Scientific Publishing Company, [2020] |
    Includes bibliographical references and index.
Identifiers: LCCN 2020015390 | ISBN 9789811219399 (hardcover) |
    ISBN 9789811219955 (paperback) | ISBN 9789811219405 (ebook) |
    ISBN 978981121941-2 (ebook other)
Subjects: LCSH: Quantum theory--Textbooks.
Classification: LCC QC174.12 .D376 2020 | DDC 530.12--dc23
LC record available at https://lccn.loc.gov/2020015390

British Library Cataloguing-in-Publication Data
A catalogue record for this book is available from the British Library.

For any available supplementary material, please visit
https://www.worldscientific.com/worldscibooks/10.1142/11801#t=suppl

Printed in Singapore

Preface

This book arose out of a one-semester class for non-science majors that I taught at Tulane University in Fall 2016. None of the students had taken any college level physics, and some had studied no physics even in high school. While it is fair to say that the course challenged the students, all 33 of them passed the class, thereby demonstrating their adequate grasp of quantum theory. I therefore know from experience that anyone can learn quantum theory with the proper guidance. And so I have written this book for all who wish to learn this subject, to be used either as a college textbook or by individual readers who wish to improve their own understanding of physics.

I owe a special debt of gratitude to Dr. Jim McGuire, who as Chair of the physics department of Tulane invited me to do a doctorate in physics, to Dr. Lev Kaplan, my doctoral advisor, whose cheerful patience made every one of our meetings a delight, and to Dr. John Perdew, who as my quantum mechanics professor inspired a deep love for the subject in me. It is seldom that one meets such brilliant physicists who are also men of genuine compassion with a total commitment to their students. It is a matter of pride and honor for me to call them my friends.

I am particularly grateful to my wife Shyla who gave her full support to my decision to pursue a career in science and embraced the challenges that accompanied that decision. It is with much love that I dedicate this book to Shyla and to our three children Melinda, Jessica and Peter who have accompanied me, whether they like it or not, in my quest for the ultimate realities of the universe.

Basil S. Davis
New Orleans, 2020.

Contents

Chapter 1

Introduction

1.1 A new understanding of reality

There have been some major upheavals in the history of science, when humans were forced to rethink their deeply entrenched ideas of reality, with the result that a prevailing worldview had to make room for a new emerging worldview. Such upheavals — or paradigm shifts — include Copernicus' heliocentric solar system, Newton's discovery that the laws of gravity and motion are the same on the earth as on the distant stars, Maxwell's insight into the nature of light as an electromagnetic wave, Darwin's theory of evolution according to which humans and animals have a shared origin, and Einstein's Relativity that brought together space, time and gravitation. Some of these discoveries provoked great consternation, since they conflicted with dominant philosophical and religious worldviews. Others were controversial within the scientific community, but with the growing success of the scientific method — according to which theories must be tested by experiment — controversies gradually yielded to consensus.

There is one significant discovery, not mentioned in the foregoing paragraph, which stands apart from all the others. The genesis of quantum theory represents the greatest intellectual leap that science has taken in the history of human civilization.

By the end of the nineteenth century, science had scored a decisive victory as a means of defining a reality that could be agreed upon by everyone regardless of their spiritual, religious or metaphysical traditions. Science had established the material universe with its component objects as an absolute reality that was independent of every subjective belief and perception. Reality was out there. Our scientific observations did not create reality, but only served to confirm the reality that was in existence whether

1

we chose to notice it or not. A tree that fell in a forest made a sound regardless of whether anyone heard it.

A falling tree makes a sound even if no one hears it

Observation could not create reality. Nor could observation alter reality. And reality in the study of physics included every physical property of a material object. So, for a particle in motion, reality included its mass, its velocity (both magnitude and direction) and its position at every instant of time — which we call the *trajectory* of the particle. While an actual physical observation might interfere with what is being observed, it was believed that in principle one could make the instruments so sensitive that the measurement would provide an arbitrarily accurate map of every single detail of the concrete reality that was already out there. This was the triumph of nineteenth century physics. Then quantum theory came along and yanked the rug from under this long drawn out victory that science had won.

1.2   A theory of particles and fields

But quantum theory does not deny reality. Far from it. Indeed, a century after the birth of this theory it is now accepted as the most accurate and most detailed description of reality that human intelligence has developed. Quantum theory does not claim that mass, energy, momentum, electric charge and other observable quantities are illusions. But it does assert that every measurement is necessarily an interaction between the object that is being studied and the subject that is doing the studying, and that in this interaction there is a limit to the information that can be obtained by the subject from the object. And this limit is not due to any imperfections in the experimental apparatus that is currently available, but is due to the very nature of reality.

Many of the rules of quantum theory run counter to what we might expect on the basis of intuition. For example, quantum theory says there is no such thing as a trajectory of a particle. Particles move from point to point in space without having an intervening path or trajectory, something that was unheard of in physics prior to the twentieth century. Trajectories are meaningful only for large objects such as baseballs, bullets and planets, all which contain billions of microscopic particles. But the individual microscopic particles themselves do not behave the same way as enormous aggregates of particles do. Indeed, the most accurate statement we can make about these particles is that they do not even exist until they are detected!

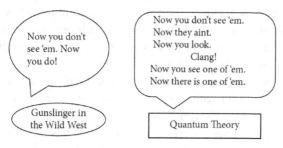

Observation of guns in the Old West and particles in Quantum Theory

Microscopic particles have no color and even the idea of shape has little meaning in this realm. And because they have no trajectories, it becomes impossible to picture them in our minds. And so quantum theory is necessarily *abstract*. This abstraction arises from the very nature of the behavior of microscopic objects. And in the realm of physics the language of abstraction has a mathematical shape. The mathematical methods used in quantum theory constitute what is known as *quantum mechanics*. In this introductory book we will not study quantum mechanics in detail, though we shall touch upon some of the basic principles of quantum mechanics. The bibliography at the end of this book lists some excellent books for the reader who wishes to move on to a rigorous study of the subject.

Quantum physics has important applications such as the photoelectric cell, positron electron tomography, the electron microscope, superconducting magnets, etc. But in this book we shall focus on the theory, with at most a cursory mention of important applications.

Quantum theory can be described as a study of particles and fields.[1] Particles and fields were both studied by classical physics. In classical physics these two entities were studied distinctly. So we had one sort of physics that described particles, and another sort of physics that described fields. Newton's Laws of motion describe particle motion, and Maxwell's equations describe the electromagnetic field. But in quantum theory both these entities come together. A particle is a particle of a field, and a field is a field of a particle. This statement may sound meaningless, but it expresses a very basic concept of quantum theory. But before we attempt to unpack this concept, we shall perform a review of classical physics (Chs. 2, 3 and 4) to get a better understanding of particles and fields in their own right.

The word *particle* means a solid object that is sufficiently small so that its dimensions can be neglected, and our interest is only in the motion of the particle. A particle could be physically as small as an electron, but if we are considering the dynamics of the solar system as a whole, an entire planet can be considered as a particle, since the size of the planet is small compared to the radius of its orbit round the sun. Particles feature prominently in the mechanics based on Newton's laws of motion.

The word *field* signifies a region of space that has an effect on certain particles that are present in that region. An apple that hangs from a tree experiences a force caused by the earth's gravitational field. A charged particle such as an electron is affected by electric and magnetic fields. A positively charged proton creates an electric field that extends in all directions leading away from the proton. This field gets progressively weaker as one moves further away. If another charged particle — say an electron — were to be brought near the proton, it would experience a force of attraction towards the proton. So in this case the proton is like the earth which produces the field, and the electron is like the apple which experiences a force due to the field. (For the sake of accuracy we clarify that both the earth and the apple generate their own gravitational fields, and both the proton and the electron generate their own electric fields.)

---

[1] In older textbooks quantum theory was commonly described as a study of particles and *waves*. While there are good historical reasons for this description, in the calculations of quantum theory today we deal mostly with abstract fields and observable particles.

1.3 Outline of the book

We will first do a quick review of the classical physics of particles which will give us a lead into the notion of particles in quantum theory. Chapter 2 will cover the fundamental classical mechanics of particle motion. It is impossible to understand modern physics without a basic knowledge of classical Newtonian physics. Chapter 3 will examine the significance of the atomic structure of matter. This chapter is entitled *Statistical Mechanics* to draw attention to this very important area of physics. Statistical mechanics played a historical role in the advent of quantum theory, as will be explained in Ch. 5, and today figures prominently in the rapidly growing branch of physics called quantum information theory. Chapter 4 will outline the classical physics of fields — specifically gravitational and electromagnetic — which will set up the background for understanding the notion of a field in quantum theory. The subsequent chapters will explain quantum theory in detail, beginning with its historical origin in Ch. 5. An important concept in modern quantum theory is entanglement, which cannot be appreciated without a knowledge of Relativity, and so Chs. 8 and 9 are devoted to explaining the concepts and the implications of Einstein's Special Theory of Relativity. We keep the mathematics to a bare minimum throughout, because this book is meant to serve as an introduction that is accessible to readers who may not have a strong mathematical background.

Quantum *mechanics* is a thoroughly mathematical discipline, requiring algebra and calculus. But because our aim is to avoid mathematics as far as that is possible, we will not do much quantum mechanics in this book. Nevertheless, the foundational principles of quantum mechanics can be learnt without a great deal of mathematics, and so we will present these principles in the course of this book.

Numerical calculations are helpful for gaining a better understanding of physics, and so we have provided simple numerical exercises throughout the book. A knowledge of basic algebra is sufficient to work out these exercises. Answers to the exercises (where required) are provided in an appendix at the end of the book.

bsd

Chapter 2

Newtonian Physics

2.1 Observation of the night sky

The earliest astronomers spent several hours night after night observing the objects in the sky. They were patient and diligent, and made careful notes of their observations. They were true scientists in the modern sense of the term. This means they relied only on their observation in order to draw conclusions regarding the reality of objects in the visible universe, rather than the authority of priests or sacred books.

They observed that the sky was filled with different luminous objects that rose in the east and traveled slowly towards the west before disappearing below the horizon. These celestial objects included the sun, the moon, the stars, and a few star-like spots of light of differing brightness. The stars seemed to be fixed to an invisible hollow sphere, and it was as though this sphere rotated around the earth. The axis of rotation appears to pass through a star in the north which we call the North Star or the Pole Star. The sun and the moon and the other star-like objects seem to revolve round the earth with different speeds. There were five such star-like lights visible at different times in the night sky: Mercury, Venus, Mars, Jupiter and Saturn. These star-like objects — together with the sun and the moon — were called Wanderers or Planets (from the Greek for wanderer).

The ancient Greek philosopher Aristarchus (died 230 BCE) suggested that instead of imagining all these heavenly objects to be orbiting the earth, it is easier to imagine the earth rotating about its axis. But this idea was so revolutionary that it was rejected. It was hard to understand how such a massive object as the earth could be set in motion, and how birds which flew above the earth were not left behind due to the rotation of the earth. Moreover, a stone thrown vertically upwards came back vertically downwards, which seemed to suggest a static earth. Many discoveries had to be made before all these objections could be satisfactorily answered.

2.2   Measurement of time

The seven-day week is probably traceable to the seven Planets seen in the sky.[1] Each day of the week was dedicated to a particular one of these heavenly lights. We can see echoes of this in our English names: Saturday for Saturn's day, Sunday for the sun's day and Monday for the moon's day. We also find this in the French names for the other days of the week, with the suffix *di* for *day*: Tuesday *mardi* (Mars), Wednesday *mercredi* (Mercury), Thursday *jeudi* (Jove = Jupiter) and Friday *vendredi* (Venus).

The pattern of the stars in the night sky changes gradually from night to night, but is repeated after a period of 365 days, or more accurately 365 and a quarter days. This was established by the ancient Egyptians who painstaking recorded the position of the bright star Sirius shortly after sunset night after night and found that Sirius returned to its original position after 365.25 days. This period of about 365 days also matched the climatic cycle of the seasons, which includes average temperatures, positions of the sun in the sky at noon, etc. So the idea that there is a natural period of time which we call a year was established in ancient Egypt, long before the year was associated with the earth's motion round the sun.

The Egyptians also divided the day into 12 equal periods and the night into 12 equal periods. Each period was called an hour. Clearly, a day hour was in general different from a night hour. Moreover, the actual length of an hour also varied with the season, day hours being shorter in winter and longer in the summer, with the reverse holding for night hours.[2] Later the hour was defined as a 24th part of a day-night duration, making it a constant measure of length at any time of day or night and all through the year. Today the scientific unit of time is the second, which is defined by an atomic clock.

As the moon orbits the earth its phase as seen from earth changes from new moon to crescent moon to half moon to gibbous moon to full moon and all the way back to new moon. So the lunar cycle also defines a period of time. This period was called a month, the word *month* being related

---

[1]I use the word Planet with a capital P to denote the term as it was used in antiquity — meaning the five visible planets plus the sun and the moon.

[2]The day/night difference and the seasonal variations would be greater in regions further removed from the equator than Egypt.

to the word *moon*. This is not an invariant period, because the length of a lunar month varies slightly from cycle to cycle. The average length of a lunar month is a little less than 30 days. So there are 12 lunar months in a year, with a few extra days left over. These days are added (unevenly) to the months to make the sum of the days of the 12 months equal to 365. Now, since the year is closer to 365.25 days, if we were to limit the year to 365 days, after 4 years we would be behind by 1 day. So 1 day is added to the year every four years (leap year). If the last two digits of the year is a multiple of 4 then it is a leap year. But even this arrangement is not perfect. Every now and then a small further adjustment is needed. So the rule is that if the last two digits are 0, then we look at the next two digits and if this number is divisible by 4, then it is a leap year. If not, it is not a leap year. So 1900 was not a leap year, and 2100 will not be a leap year. But 2000 was a leap year, and the next leap year ending in two zeros will be 2400.

Exercise 2.1.

(a) How many seconds are there in a month of 30 days?

(b) How many seconds are there in a year (365.25 days)?

2.3 Ptolemy's model

Since the sun presents a circular face it seemed natural to assume that the sun is actually a sphere. Likewise, although the moon goes through phases, it made sense to assume that the moon is a sphere and the dark part of the moon represents the night regions. The shapes of the other planets could not be discerned with the human eye, but it seemed logical to assume they too are spheres. What about the earth? The earth looks more or less flat, but the early scientists figured out that the earth is actually a sphere.

The Greek philosophers came to this conclusion through careful observation and logical deduction. Chief among their observations were the following:

1. A ship dips below the horizon when it travels far from the shore, showing that the ocean surface is curved.

2. The noon day sun is seen further south in the sky in more northern climates.

3. The shadow of the earth on the moon's surface during a lunar eclipse is circular.

So the earth and all the planets were known to be spherical. The sphere was therefore a perfect three-dimensional shape. It made sense to say that the circle was the perfect two-dimensional shape. This had major consequences for understanding the movement of the solar system.

Ptolemy (died 178 CE) was an Egyptian astronomer who studied the celestial objects with great precision. He concluded that the heavenly bodies all travel around the earth in circles — with each circle being called a deferent — with somewhat differing speeds. With the exception of the sun, the moon, and the stars, the celestial objects (i.e. the five planets) also move in a cyclic path — called an epicycle — about the main circular orbit or deferent. This dual motion was necessary to explain why these planets seem to backtrack their path in the sky every now and then instead of progressively moving along in a single direction.

Ptolemy's model prevailed for over a thousand years and was the dominant model of the solar system used by scientists until it was seriously challenged by Copernicus.

2.4 The Copernican revolution

Copernicus (died 1543) showed that it is easier to explain all the astronomical data of the movements of the heavenly bodies by assuming that the earth rotates on its axis and that it goes round the sun and that all the planets also orbit the sun, except that the moon orbits the earth and travels along with the earth in its orbit of the sun. Copernicus' model is called the heliocentric — sun at the center — model, as opposed to the geocentric — earth at the center — model of Ptolemy. Because Copernicus' theory brought about a major shift in our most basic understanding of the visible universe, any major change of generally accepted ideas has come to be called a *Copernican shift*.

But the Copernican revolution did not happen overnight. Copernicus was fully aware of the controversy that his theory would create, and so he published his thesis only on his deathbed. He was condemned by the Christian leaders of Europe, both Catholic and Protestant, on the charge that his model contradicted the Bible. Though the Bible has not changed, today no Christian leader claims any contradiction between the Bible and the Copernican model of the solar system. But the religious objection was not the most serious obstacle to the universal acceptance of Copernicus' theory. Let us recall that the philosopher Aristarchus (died 230 BCE) had put forward a heliocentric model of the solar system several centuries earlier.

The opposition to Aristarchus certainly did not come from religion but from observation. The clouds, the air in general, and creatures that flew through the air all acted as though the earth were stationary. It required Newton to explain how these phenomena could be consistent with a rotating earth.

After Copernicus, astronomers abandoned Ptolemy's model and made further observations that both confirmed and refined Copernicus' model.

Tycho Brahe (died 1601) constructed an extremely accurate observatory. He recorded the position (i.e. the angular position of a celestial object when viewed from a point on the ground) of the heavenly bodies at different times over a period of several years.

Johannes Kepler (died 1630) made use of Brahe's data and found that they conflicted slightly with the data used in support of Ptolemy's and Copernicus' models. Kepler found that Brahe's data could be explained by suggesting that the earth and the planets did not travel round the sun in perfect circles, but rather in ellipses (oval paths), with the sun at one focus of the ellipse.

With the invention of the telescope, Galileo Galilei (died 1642) was able to show that the planet Venus showed all the phases — full, gibbous, half, crescent — as the moon. This was perfectly explainable on Copernicus' heliocentric model, but not on Ptolemy's geocentric model. Ptolemy's model did not allow for a full phase of Venus to be visible from earth. Thus there was unquestionable evidence — in addition to Brahe's observations and the calculations of Kepler — that the planets do indeed orbit the sun.

A year after Galileo's death saw the birth of Isaac Newton who became the father of what we today call Physics in the modern sense of the term. Newton used the theories of Kepler and the observations of Galileo to establish a mathematical description of physics which still endures today, though it has been greatly modified by Albert Einstein's theory of Relativity and the Quantum theory developed by several physicists of the early twentieth century.

2.5 Newton's laws

Several Greek philosophers made significant contributions to our understanding of nature, but it was Aristotle who drew up a system for understanding fundamental physical phenomena. Aristotle believed — as did most Greek philosophers — that all matter was composed of the elements: earth, air, water, fire and ether. Vertical motion was natural. Solid objects fell to the ground because they sought to be close to the earth, since they were also made of earth. Liquids fell or flowed downwards because they sought to be united to the seas, the domain of the element called water. Flames shot upwards because the domain of fire was above. The purpose of all natural motion was goal oriented. Horizontal motion, on the other hand, was violent, and needed an agent to sustain it. The motion of the stars was circular, because this was the property of ether.

But Aristotelian physics was not based on carefully controlled observation. Laboratory experiments showed that bodies had this property called inertia which would cause a moving body to continue its motion in the absence of friction or air resistance.

Newton raised the concept of inertia to the status of a law of physics. He was thus able to show that all motion could be explained on the basis of his three laws of motion:

First Law:
Every object remains at rest, or moves with a constant speed in a straight line, unless compelled to do otherwise by an external force.

As the earth rotates, the air that is close to the ground also moves with the same speed as the ground. Every object that floats or flies through the air is also carried along with the motion of the earth. Inertia keeps everything going together. So we cannot tell from an observation of the atmosphere that the earth is in motion, any more than we can tell that an airplane is moving by observing an object that is dropped inside the plane. Of course, because the motion of the earth is a rotation and not a simple translation along a straight line, there are other factors that give rise to atmospheric phenomena such as hurricanes, and it is these that reveal the rotation of the earth.

Second Law:
An external force $\mathbf{F}$ applied to an object of mass $\mathbf{m}$ would impart an acceleration $\mathbf{a}$ to the object in the direction of the force such that $\mathbf{F} = \mathbf{ma}$.

Acceleration is not just increase of speed, but any change of velocity in magnitude or direction. So an object moving on a circular path is accelerating because its direction of motion is changing. It can be shown with some calculation that an object moving along a circle of radius r with a constant speed v experiences an acceleration of magnitude v^2/r directed towards the center of the circle.

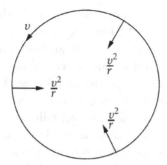

A particle moving with constant speed v along a circle of radius r experiences an acceleration of magnitude $\frac{v^2}{r}$ towards the center of the circle.

An important quantity in the study of motion is the *momentum* of an object, written as p and defined as the product of mass and velocity: $p = mv$. Acceleration is the rate of change of velocity. So the product ma is the rate of change of momentum. So Newton's Second Law can be stated as: Force = Rate of change of momentum.

Exercise 2.2.

(a) A car of mass 1500 kg accelerates from rest to a speed of 60 kmph in 10 seconds. Find the average force applied on the car by the engine. (Convert kmph into meters per second. Remember, average acceleration = change of velocity divided by time.)

(b) Taking the distance of the earth to the sun as 150,000,000 km find the circumference of the earth's orbit in meters. Hence find the speed of the earth as it travels through space. Use the result of Exercise 2.1 (b).

(c) Find the acceleration of the earth as it orbits the sun.

Third Law:

When an object A applies a force on an object B, the object B simultaneously applies an equal and opposite force on object A.

The Third Law explains why we do not go through the floor. Our weight applies a force on the floor, and the floor applies an equal and opposite force — called the *normal force* — on our feet.

These laws have successfully explained horizontal motion. What about vertical motion? According to the Second Law an accelerating body must be driven by a force. So if an apple accelerates to the ground it is because there is a force acting upon it. Moreover, this law also states that the force is proportional to the acceleration. Not all bodies are pulled to the earth by the same force. Some objects are heavier than others. But heavier and lighter objects fall at the same acceleration in the absence of air resistance. The Second Law can also be written as $a = F/m$. The acceleration equals the force divided by the mass. Since all falling objects accelerate at the same rate in a vacuum where there is no air resistance, Newton's Second Law indicates that there must be a force — the force of gravity — acting on an object that is proportional to the mass of the object. The greater the mass, the greater the force, and so the ratio of force to mass stays constant.

So the force exerted by the earth on an apple is proportional to the mass of the apple. Newton's Third Law requires that if the earth exerts a force on the apple, the apple must exert an equal and opposite force on the earth. Through an argument from symmetry it follows that the force exerted by the apple on the earth should be proportional to the mass of the earth. So this suggests that the mutual force of attraction between two objects should be proportional to the product of the masses of the two objects.

$$F \propto m_1 m_2$$

Newton extended this law to include all the heavenly bodies such as the moon, the sun and the planets. What about circular motion? It is not ethereal motion, but a form of acceleration, as we have seen. And acceleration requires force. If we tie a stone to a string and twirl it around, the stone will fly in a circular path because the tension in the string provides the force that generates the acceleration (equal to v^2/r where v is the speed of the stone and r the length of the string). In the case of a planet circling the sun, the force comes from gravity. Using Kepler's data of planetary motion Newton inferred that the further a planet is from the sun, the smaller is its acceleration, and that this acceleration is inversely proportional to the square of the distance of the planet from the sun. So Newton was able to conclude that the force between two objects of masses m_1 and m_2 at a distance of r from each other is proportional to the product $m_1 m_2$ and inversely proportional to r^2. So this proportionality can be written as

$$F \propto \frac{m_1 m_2}{r^2}$$

A proportionality relation can be written as an equation by introducing a constant factor. So Newton's Law of Gravitation is fully written as

$$F = G\frac{m_1 m_2}{r^2} \qquad (2.1)$$

where G is called the Universal Gravitational Constant. Scientists use the *Système Internationale* or SI system for defining the units of measurement. This system is also called the MKS system because its units are meter, kilogram and second.[3] $G = 6.67 \times 10^{-11}$ in SI units.

What is the gravitational force acting on an object of mass m near the surface of the earth? Let us call the mass of the earth M. Let the radius of the earth be R. Consider an object a short distance above the ground, where by short distance we mean not more than about 10 kilometers. This is small compared to the radius of the earth which is about 6370 km. The force exerted by the earth on this object of mass m is therefore

$$F = G\frac{mM}{R^2}$$

where the distance between the mass and the earth is taken to be the distance between the mass and the center of the earth, which is R.

If this object were dropped from a small height above the ground, it would fall with an acceleration which we shall call g that is given by Newton's Second Law as $F = mg$. Therefore

$$mg = G\frac{mM}{R^2}$$

Cancelling m from both sides, we obtain

$$g = G\frac{M}{R^2} \qquad (2.2)$$

The mass of the earth $M = 5.97 \times 10^{24}$ kg, the radius of the earth $R = 6.37 \times 10^6$ m and $G = 6.67 \times 10^{-11}$ in SI units. Putting these numbers into the equation, we obtain the acceleration g of a falling object to be 9.8 m/s². So the force of gravity acting on an object of mass m is given by mg. We call this force the *weight* of the object in a technical sense. So a mass of 10 kg would have a weight of 98 newtons (written 98 N) where the newton is the SI unit of force.

---

[3]The United States continues to use the outdated British FPS system. This is an example of systemic inertia.

Exercise 2.3.

(a) Earlier you found the acceleration of the earth as it orbits the sun. Knowing the distance of the earth to the sun and the value of G find the mass of the sun. (Hint: See Eq. (2.2). Here g is the acceleration of an object in the gravitational field of the earth. Replace this by the acceleration of the earth in the gravitational field of the sun.)

(b) The mass of the moon is 7.35×10^{22} kg and the radius of the moon is 1726 km. Find the acceleration with which an object would fall due to the force of gravity close to the surface of the moon.

(c) If an object is fired horizontally close to the ground it could go into orbit round the earth if the speed of the object is sufficiently high. Using the fact that the centripetal acceleration of an object moving along a circle is v^2/r find the minimum horizontal velocity v with which a rocket must be fired for it to go into stable orbit round the earth. For simplicity assume the earth is a smooth sphere with no hills, trees, buildings or birds, and ignore air resistance.

2.6 Work and energy

When a force acts on an object and displaces the object through a distance, we say that *the force performs work on the object*. If the displacement is exactly along the direction of the force, the work done is simply the product of the force and the displacement. If the force is at an angle to the displacement, the work done is less than the product of force and displacement. If the force is at right angles to the displacement, the work done is zero. Mathematically, we say work $W = FS \cos\theta$ where θ is the angle between the force of magnitude F and the displacement of magnitude S.[4]

When an apple falls from a tree, work is done *by* the force of gravity *on* the apple. Because the force of gravity is in the same direction as the

---

[4]Consider a triangle with one right angle = 90^0. The side opposite to this right angle is called the *hypotenuse*. If θ is one of the other two angles, then the side between the 90^0 angle and the angle θ is called the adjacent side of the angle θ and the third side is called the opposite side. We define $\cos\theta$ as the number obtained by dividing the length of the adjacent side by the length of the hypotenuse.

displacement the work done is positive. If the apple drops through a vertical height h, the work done by gravity on the apple is mgh.

What is the effect of this work done by the force of gravity? The immediate effect is that the apple speeds up. A moving object is said to have kinetic energy by virtue of its motion. If the mass is m and the speed is v the kinetic energy of the object is given by

$$KE = \frac{1}{2}mv^2$$

Prior to the apple's fall it was at some height above the ground. If this height is h meters, then we say that it had a potential energy of mgh. As it falls, its potential energy changes into kinetic energy and work is being done on the apple by the force of gravity.

The *Work-Kinetic Energy Theorem* states that the work done by a force on a body is equal to the change of kinetic energy of the body.

Potential and kinetic energy are forms of mechanical energy. We see that potential energy can be converted into kinetic energy. The reverse happens when we throw an object upwards. In the next chapter we shall see that there is another kind of energy called heat energy, and that it is possible to convert heat into mechanical energy and vice versa. So there is a universal law — called the Law of Conservation of Energy — which states that energy cannot be created or destroyed, but may be converted from one form to another. In SI units energy is measured in joules (J).

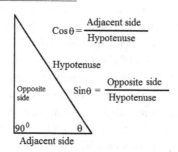

The ratio $\sin\theta$ is defined as shown.

Exercise 2.4.

(a) What is the kinetic energy of the earth as it orbits the sun, ignoring the kinetic energy due to the rotation of the earth about its axis?

(b) 50,000 kg of water drop every second down a waterfall through a height of 30 m. If 80 per cent of this gravitational energy could be converted to electricity, how much electrical energy can be produced at the bottom of the fall per second? Energy produced per second is called *power* and expressed in units called watts (W).

2.7 Determinism

A major consequence of Newtonian physics was that it created a sense of absolute universalism in the minds of many people. According to the laws of Newton every object follows a deterministic behavior. For example, consider two spheres moving along a straight line towards each other and moving away from each other along the same straight line after collision. If we know the masses and velocites of the spheres before they collide, we can predict the velocities of the spheres after they collide, assuming the total kinetic energy to be unchanged in the collision. In the following chapter we shall see how we came to know that all matter is made of atoms and molecules. If it were possible to measure the masses, positions and velocities of every single atom and molecule in the universe at a given moment of time, then it is theoretically possible to predict the exact configuration of all the atoms and molecules an hour later, a day later, a year later, a billion years later. This means that the exact state of the universe a billion years from now has already been determined. Nothing can change the flow of the history of the universe. Whether you will remain alive and if so what color clothing you will be wearing on a day exactly 10 years from today has already been determined. There is no such thing as free will because the circuits of the brain follow the laws of physics and whatever they do is dictated by necessity and they cannot do anything other than what is determined by the laws. So free will is an illusion. When we think we are choosing something over another we are simply moving in the direction determined by the laws of physics. Ironically, this sort of scientific determinism leads to conclusions very similar to nonscientific beliefs in astrology and fate.

With the advent of the quantum theory and relativity in the early years of the 20th century, Newtonian physics was replaced by modern physics which basically states that Newtonian principles need to be revised when we consider very small objects such as atoms and very fast moving objects such as light. Modern physics does not posit a deterministic universe. It is intrinsically impossible to make an exact measurement of all the quantities such as position, velocity, etc. of even a single atom, leave alone billions of atoms. Not only is precise measurement impossible, but prediction is also ruled out by the laws of quantum mechanics. Thus determinism has collapsed. One can no longer use Newtonian physics to argue against the possibility of free will. The philosophical debates regarding free will became far more sophisticated as a result of quantum theory.

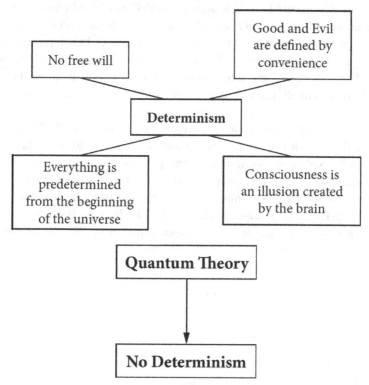

2.8 Summary

The heavenly bodies appear to be moving in a circular motion round the earth. But the cumulative results of painstaking observations made by astronomers through the centuries enabled Copernicus to conclude that the sun and not the earth is the center of the solar system. But if the earth also moved in a circular orbit, then one could no longer accept the ancient belief that circular motion was a special property of the heavenly bodies.

The Copernican worldview required a scientific explanation for this perpetual motion of the planets round the sun. Newton was the first scientist to explain the kinematics of both celestial and terrestrial objects. He provided a simple mathematical system that explained all sorts of motion. The force of gravity attracts all objects to each other. The force between two objects is proportional to the product of the masses and inversely proportional to the square of the distance between them.

In addition to the law of gravity Newton also enunciated the three laws of motion. All objects have this property called inertia, whereby they tend to remain at rest or move in a straight line with constant speed. In order to change this state, an external force has to be applied. This force is numerically equal to the product of the mass of the object and its acceleration under this force. Moreover, when one object applies a force on another, the second object simultaneously applies an equal and opposite force on the first.

Newton's laws of motion and gravity form the basis of all classical mechanics, the branch of physics dealing with motion and forces. These laws successfully explained the motion of the planets as well as all motion on earth. They seemed to be absolutely infallible rules followed by all matter in the universe. This led many people to believe that the universe is deterministic. Everything happens the way it does because it cannot happen any other way. All things have been predetermined by the laws of motion.

Chapter 3

Statistical Mechanics

3.1 Atoms

The Greek philosopher Democritus (460–370 BCE) proposed that material objects cannot be divided into smaller and smaller pieces *ad infinitum*, but that a stage would be reached when it would be impossible to divide matter any further. Thus all matter was made of very tiny grains which were so small that they could not be seen. Democritus claimed that these tiny grains could not be divided any further, and so he called such a grain an atom, which in Greek means "indivisible".

What evidence is there for the existence of atoms?

Democritus argued that odors and aromas are caused by atoms breaking off from the material and flying through the air and eventually entering our nostrils. This explanation is valid even today, though we would use the word "molecule" rather than atom to describe the smallest particles of the substance that we are smelling.

The first visible proof of Democritus' hypothesis of these atoms was the observation of the random erratic motion of pollen grains and saw dust grains suspended in water when seen through a powerful microscope. This motion is called Brownian motion, after the botanist R. Brown who reported his observation in 1827.[1] The pollen grains were seen dancing about in a zigzag haphazard manner. The explanation is that the tiny invisible particles of water were hitting the pollen grains from different directions at different speeds, and the pollen grains were reacting to these

---

[1] For the record, though, this phenomenon had been observed and reported earlier by Jan Ingen-Housz in 1784, but the report was so brief that it escaped the notice of subsequent researchers. Though unfortunate for the memory of Ingen-Housz, this historical lapse is perhaps to the advantage of English speakers, since it is easier to say "Brownian motion" than "Ingen-Houszian motion"!

collisions by moving in random directions. Today we know that the smallest particles of water are not really atoms, but molecules. Brownian motion provided the first visible proof that matter is made of small indivisible particles.

The atomic theory of matter is able to explain the observations of chemistry. Chemistry investigates reactions whereby elements combine with other elements to form compounds, and compounds react with other elements or compounds to produce other substances, both elements and compounds. Around 1880 the chemist John Dalton found that when elements combine to form compounds they do so in fixed proportions by weight. As an example, 1 gram of hydrogen combines with 8 grams of oxygen to form 9 grams of water, and 2 grams of hydrogen combine with 16 grams of oxygen to form 18 grams of water. This seems to suggest that substances are made of atoms which have fixed masses, because this observation would make sense if each substance was made of small microscopic particles all having the same mass and that the masses of these fundamental particles varied from one substance to another. So Dalton's observation offered a good indirect proof for the existence of atoms.

Another major victory scored by the atomic theory is that it has been able to explain our most basic experiences of heat and cold. Temperature is something everybody experiences. The application of heat to effect useful changes — particularly for cooking food — had been recognized since the discovery of fire. The industrial revolution took off when heat was recognized as a form of energy that could be harnessed to do useful work such as in the steam engine and the automobile engine. The study of the relationship between heat and mechanical energy is called *thermodynamics*. Physicists made considerable progress in discovering important laws of thermodynamics before they understood the nature of heat. The atomic theory finally explained that heat is nothing but the sum total of the energies of the individual atoms or molecules of a material object — solid, liquid or gas. Temperature is simply a measure of the average kinetic energy of the microscopic atoms or molecules of an object.

We shall do a quick review of the laws of thermodynamics and show how these laws make sense in the light of the atomic theory of matter.

3.2   The laws of thermodynamics

The word "thermodynamics" is a combination of "thermo" meaning *heat* and "dynamics" meaning the effects of *forces*.

So in the study of thermodynamics we are interested in the interplay between heat and *work*. Let us recall that we learned earlier (Ch. 2) that work is done when a force causes a body to undergo a displacement.

When a dynamo is made to rotate swiftly, it produces electricity. One way of causing a dynamo to rotate is to attach paddles to it and place it under a waterfall. The falling water has mechanical energy due to the speed with which it falls from the upper level. We say that the water has potential energy by virtue of its elevation above the ground and that this potential energy is converted to kinetic energy as the water gathers speed in its descent. Both potential and kinetic energies are forms of mechanical energy. So in this example — known as hydroelectricity — the mechanical energy of the water is converted to mechanical energy of the rotating dynamo which in turn generates the electricity. Thus mechanical energy is converted to electrical energy.

If a waterfall is not available, but there is fuel available for burning, such as coal, gasoline or nuclear fuel, then these fuels can be burned and heat energy obtained. This heat energy can be used to drive a dynamo which would then produce electricity. So in this example — known as thermoelectricity — heat energy from the burning fuel is converted to mechanical energy in the rotating dynamo which then generates the electricity.

The obvious disadvantage of thermoelectricity is that the fuels will eventually be exhausted at some time in the future. There is also another disadvantage, and that is, in converting heat into work there is always a waste of heat energy. Not all the available heat can be converted into useful energy. And this follows from the laws of thermodynamics, which, as we shall see presently, follow from the atomic or particle nature of matter. We now turn to these laws.

The study of thermodynamics is classified under four laws: the zeroth law, the first law, the second law and the third law.

Brief explanations of these laws are laid out below:

Zeroth Law:

The zeroth law states that if two bodies are separately in thermal equilibrium with a third body, then they must be in thermal equilibrium with one another. By thermal equilibrium we mean simply that no heat flows between the two bodies. This law enables us to define the concept of temperature. Two bodies that are in thermal equilibrium with one another have the same temperature. So if body A and body B have the same temperature, and body A and body C have the same temperature, then by the zeroth law, body B and body C must have the same temperature. This means we can define temperature as an absolute quantity, regardless of the nature of the body that has the temperature. So thermometers can be built to measure the temperatures of objects having compositions totally different from that of the thermometer. We express temperature as a number. This number is commonly written using one of two common scales, Celsius and Fahrenheit. These scales are related by the formula

$$\frac{C}{5} = \frac{F - 32}{9}$$

According to this formula 100^0 C $= 212^0$ F, which is the temperature at which water boils under normal atmospheric pressure.

Physicists who worked on the behavior of gases as they expanded with increase of temperature noticed that there seemed to be a lowest possible temperature which is -273.15^0 C. This is called Absolute Zero. One could think of Absolute Zero as the temperature of a body after all its heat has been taken away. So there is a lower limit to temperature. Since it is impossible to get colder than Absolute Zero, it makes sense to define a scale of temperature in which 0 corresponds to Absolute Zero. Such a scale is called the Kelvin or Absolute Scale.[2]

Exercise 3.1.
Which is colder: (a) 0^0 C or 0^0 F? (b) -20^0 C or -20^0 F?
(c) -40^0 C or -40^0 F? (d) -60^0 C or -60^0 F?

First Law:
The first law is a statement of the conservation of energy. It states that when heat energy is given to a body, part of it goes to raise the temperature, and hence the internal energy of the body increases, and the rest goes to do external work, which is done by the expansion of the body. Since this is a law of conservation of energy, it can also be applied to a case when work is done on a gas by compressing it. In this case the gas would get heated up, and may give up some of its heat to the surroundings. Here also there is a balance of energy. No energy is created and no energy is destroyed.

If the amount of heat energy supplied to a body is ΔQ, the rise of internal energy of the body ΔU, and the external work done by the body ΔW, then the first law can be expressed as

$$\Delta Q = \Delta U + \Delta W$$

---

[2]The lowest temperature on this scale is 0 K, the melting point of ice is 273.15 K and the boiling point of water is 373.15 K. Notice that we do not put the degree symbol ^{0} in the Kelvin scale.

Second Law:

The second law is a statement of irreversibility. It can be stated in many different ways. The simplest statement is that heat always flows naturally from a hotter to a cooler body.

Both the first and the second laws deal with the conversion of heat into work and work into heat. The first law tells us that heat and work are different forms of energy and that one can be converted into the other. The second law places restrictions on the conversion of heat energy into work. Heat and work (mechanical energy) are not reversible. Whereas mechanical energy can be converted entirely into heat energy, the reverse cannot take place. For example, when a meteor falls through the atmosphere, it has a large kinetic energy because of its speed, and this kinetic energy is entirely converted to heat energy as the meteor burns up in the atmosphere. When brakes are applied to a moving car, the kinetic energy of the car is entirely converted to heat energy in the wheels and the road. But the heat energy that is so generated in either of these examples cannot be converted back to kinetic energy.

Third Law:

The third law states that it is impossible to cool a body right down to absolute zero (0 K), even though it is theoretically possible to come closer to this temperature with each attempt.

All the laws of thermodynamics can be fully explained on the atomic or molecular theory of matter. Heat is a manifestation of the energies of the molecules of a body — the sum of all the kinetic and potential energies of all the molecules. Here the forces that give rise to the potential energy are not due to gravity but due to attraction and repulsion between molecules. The phenomenon of Brownian motion showed that the motion of the molecules in a liquid is erratic and random. The molecules move in all possible directions with a range of velocities that change in magnitude and direction each time a molecule collides with another or with the molecules of the walls of the container. Thus it is futile to try and follow the movements of any one molecule. The best we can do is to investigate the overall aggregate or statistical behavior of these molecules. The study of the behavior of matter in terms of the collective motion of the molecules is therefore called *statistical mechanics.*

3.3   Statistical mechanics

Every substance — whether an element or a compound — is made up of molecules. Each molecule contains one or more atoms. The molecule of an element contains one or more atoms of the same kind. A molecule of a compound contains two or more atoms, which are not all of the same kind. Sulfuric acid has the formula H_2SO_4, which means a molecule of sulfuric acid contains two hydrogen atoms, one sulfur atom and four oxygen atoms. Because atoms — and molecules — are so small, there is a very large number of these particles in any observable piece of matter. A measure of this large number is Avogadro's number N_A = number of molecules present in 1 mole of any element or compound. A mole is the molecular weight expressed in grams. 1 mole of hydrogen gas (H_2) has a mass of 2 grams. $N_A = 6.02 \times 10^{23}$. So 1 gram of hydrogen gas (H_2) contains about 3.01×10^{23} molecules. This means 1 hydrogen atom has a mass of about 1.7×10^{-24} grams. This is a very small quantity. 16 grams of oxygen gas (O_2) contain about 3.01×10^{23} molecules. Because of the very large number of molecules present in this small mass of oxygen, we need statistical mechanics to provide a reliable description of the observable behavior of the gas.

Exercise 3.2.
(a) The formula for water is H_2O. How many molecules are there in 1 gram of water?
(b) The molecular weight of a substance is the sum of the atomic weights of the atoms in a molecule of the substance. Given the following atomic weights: $H = 1, O = 16, S = 32$, find the molecular weight of sulfuric acid H_2SO_4.

The temperature of a body is a measure of the average kinetic energy of the molecules of the body. If two bodies are in thermal communion with each other, there will be a transfer of kinetic energy from the molecules of one body to the molecules of the other body, until both bodies have the same average kinetic energy of their molecules. This means they will have the same temperature. This is the explanation for the zeroth law. This also explains the second law, as we shall see further on.

The temperature of a gas is proportional to the average kinetic energy of its molecules. Consider a gas contained in a cylinder enclosed by a piston. If some heat is supplied to the gas, its temperature will increase, and so the molecules will have greater kinetic energy. This greater kinetic energy will mean that the molecules will pound on the piston with greater force, causing the piston to move outwards. So the gas expands and the force of this expanding gas does work on the piston. This is an illustration of the first law.

3.3.1 *One-dimensional gas*

A helium molecule which consists of a single atom can be thought of as a rigid sphere. Diatomic molecules such as hydrogen and oxygen have a dumbbell shape. Molecules with three or more atoms have more complex shapes. Let us now limit our discussion to the simplest kind of gas — one consisting of identical monatomic molecules such as Helium, Argon or Neon. Each of these molecules can be modeled as a tiny rigid sphere.

Suppose all these rigid spheres were lined up along their line of centers, i.e. like a one-dimensional array of identical billiard balls.

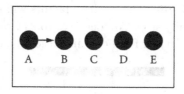

And let us say that this array of balls is suspended within a zero gravity box somewhere in outer space. If the sphere A at the far left were set in motion towards the right, it would collide with the next one, which in turn would collide with the sphere next to it, and so on till the last sphere E moves forwards, hits the wall of the container, bounces back, hits the previous ball D, which in turn hits the ball C behind it, and so on until the ball A moves to the left wall, bounces back, hits the ball B, and the process continues indefinitely. As long as all the collisions are elastic, i.e. with no loss of kinetic energy, the process will be repeated forever. Moreover, the process is also time reversible. If we were to record the motion of the balls for a period of time and play the film backwards it would be impossible to find any essential difference between the forward time and backward time sequences of motion. What we have just described is a model of a one-dimensional gas, which of course does not exist in nature. But what is important for our purposes is that a one-dimensional gas in a gravity-free environment is a time reversible system.

The educational toy called "Newton's Cradle" is an approximate illustration of this process. A typical Newton's Cradle has about five identical metal balls suspended by strings from a horizontal support. The strings all have the same length and they are spaced apart in a straight line so that when the apparatus is stationary all the balls hang vertically at equal distances from their neighbors. If now the ball at one end is pulled away from the rest and released, it would fly like a pendulum and hit the next ball. There is a total transfer of momentum from the first ball to the second with the result that the first ball becomes stationary. The momentum is then communicated from the second to the third to the fourth to the fifth. The fifth ball flies away from the remaining balls which are now all stationary. The fifth ball makes a pendulum-like swing and returns and hits the fourth ball, which communicates the momentum to the neighboring ball and so on all the way to the first ball which now pulls away from the others and moves to the left like a pendulum before swinging back and hitting the second ball and so the sequence of movements is repeated all over again. There is loss of energy at each collision, as kinetic energy is converted to sound and heat energy, and there is air resistance. These factors will cause the process to slow down and eventually stop. An ideal Newton's Cradle with zero energy dissipation would be a model for a one-dimensional gas.

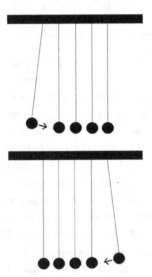

A *degree of freedom* is a particular way in which a molecule is free to move. And because a molecule in this scenario can execute only one kind of motion, which is to move along a straight line, such a molecule has a single degree of freedom.

3.3.2 *Two-dimensional gas*

Next we consider a two-dimensional gas. Again we consider a container in a gravity-free environment. Here the balls are floating at different points but their centers are all in the same plane. This time, the spheres are not all aligned in straight lines. Now, if one ball were given a push in any direction, it would hit another, which would hit another, and so on, but these collisions would not necessarily be head on collisions.

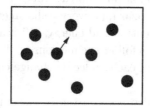

These collisions would be random. Eventually, the balls would be moving haphazardly in all directions, while remaining in the same two-dimensional plane. But the kinetic energy of the balls has now been distributed evenly along two dimensions. The random statistical nature of the motion ensures that the average kinetic energy due to the motion in any one direction equals the average kinetic energy due to motion in any other direction. Every two-dimensional motion can be resolved into motion in two mutually perpendicular directions, and we call each such perpendicular direction a *degree of freedom*. So a monatomic molecule that is capable of moving in two dimensions has two degrees of freedom. Each degree of freedom has the same average kinetic energy. The average energy per molecule has been divided equally between its two degrees of freedom. Clearly, the dynamics of this two-dimensional gas are not time reversible. In forward time the energy gets distributed evenly between the two degrees of freedom. One does not observe the reverse happening in nature. One does not see a collection of billiard balls initially moving randomly gradually changing their motion until all the balls are moving in one direction only. Thus a two-dimensional gas is an irreversible system.

3.3.3 *Three-dimensional gas*

Let us next picture a physically real cubic meter of helium gas inside a cube of sides 1 meter in a laboratory on earth. What are the molecules of helium doing? They are not stationary. If they were, they would all be lying at the bottom of the container like a lot of microscopic apples in a largely empty crate. But these molecules are constantly moving. They move fast like tiny bullets and so gravity does not play a perceptible role in their motion. As they move inside the cube they collide with the walls of the cube and change direction. They also collide with one another. With each collision, the molecules abruptly change velocity and exchange kinetic energy with each other, and with the molecules on the walls of the container. Since the collisions are entirely haphazard, at every collision each molecule undergoes a random change of momentum and energy. With about 10^{24} molecules to deal with it is impossible to follow the motions of all of them within the gas. The energy and velocity of each molecule change constantly from collision to collision.

Because these molecules are constantly exchanging energy with each other, it makes better sense to talk about averages than actual values when dealing with such large numbers. When the molecules are moving in the most random fashion, the kinetic energy gets distributed equally among the three degrees of freedom of the three-dimensional gas. This is an example of the Principle of Equipartition of Energy. This Principle is simply a statistical consequence of the random nature of the motion of a very large number of particles which are constantly exchanging kinetic energy and momentum as they collide with one another and with the walls of the container.

3.3.4 *Third law of thermodynamics*

The third law can be explained by an analogy. Suppose a moving sphere A collides head-on with an identical stationary sphere B. The first sphere will stop, and the second sphere will move with the same velocity possessed by the first sphere before the collision. The first sphere will not stop if the second had any velocity at all. Suppose A represents a gas that we are trying to cool to absolute zero. At absolute zero the kinetic energy of the molecules is zero. So if we want to reduce the temperature of a gas to zero, we must place it in contact with a gas that is *already at absolute zero, with stationary molecules.* Unless we can find such a gas somewhere in the universe — which is impossible, considering that the universe is constantly cooling down from a very hot initial state — it is impossible to reduce the temperature of any gas to absolute zero.

3.3.5 *Second law of thermodynamics*

Suppose we have a thermally insulated container of gas with two compartments, one having gas A initially at 150^0 C and the other having gas B initially at 50^0 C and these compartments are separated by a wall that permits heat to flow through it. Since heat is due to the kinetic energies of the molecules of the gas, the average kinetic energy of the molecules of A is greater than the average kinetic energy of the molecules of B. Due to the randomness of the collisions of the gas molecules with the molecules of the wall that separates them, the molecules of A will gradually impart some kinetic energy to the molecules of the wall, which in turn will pass on the kinetic energy to the molecules of B. The result is that the molecules of A will gradually slow down, and the molecules of B will gradually speed up. This transfer of kinetic energy will continue until both compartments have the same average kinetic energy. Thus both the gases will have the same temperature, and heat has flowed from the hotter body A to the cooler body B. It is impossible for the reverse to happen. And thus the Second Law of Thermodynamics follows from the fact that any piece of matter is made of a very large number of particles in random motion.

Because heat cannot flow from a cold body to a hot body by itself, the Second Law of Thermodynamics provides a unique arrow of time. Suppose an ice cube were placed in a glass of warm water. A video recording will show the ice melting as it receives heat from the water. If the video were played backwards it would show a tiny piece of ice gradually becoming bigger until it acquired the shape of a cube floating on the warm water. It is evident that this sort of time reversal cannot occur in nature. The flow of time is like the flow of heat. It cannot be reversed. As we saw earlier in this chapter, when we have a large number of microscopic particles, no matter how orderly they are arranged in the beginning, once the system is set in motion, the random collisions will create a disorder from which the original order can never be retrieved. This has some very important consequences.

One consequence is the diminishing of available energy. An array of molecules all moving together can apply a concerted force which can therefore do a lot of work on an object and impart a corresponding energy to the object. But if the molecules are moving haphazardly, the force they can exert together is considerably less, and so the amount of energy that can be provided is less. Thus, in an irreversible process the amount of available energy decreases. So the Second Law can also be stated as: natural processes always take place in such a way that the amount of available energy decreases.

Another consequence is the collapse of orderliness. An array of molecules all moving with the same velocity parallel to each other is a highly orderly system. But as the system is left to itself, the degree of orderliness will gradually diminish until there is total randomness. So the Second Law can also be stated thus: natural processes will always take place in such a way that there is a loss of order. Disorderliness is also called *entropy*. So another formulation of the Second Law: Natural processes occur in such a way that the entropy increases. Yet another consequence is the loss of information. We could create different arrays of molecules which are all orderly, but not identical with each other. Let us say we have two boxes with the same number of molecules all moving parallel to each other. In one box we divide the molecules into two parallel arrays with a gap between them. In the second box we have the same number of molecules, all parallel to each other, but without a gap. We could label the first box **0** and the second box **1**. The distinction between the two boxes allows us to store *information*. The simplest information is binary — yes or no. We could agree that **0** means yes and **1** means no, or vice versa. Now, suppose we allow both the boxes to stand for a while. After some time all the molecules in both boxes will be moving at random, and the gap between the molecules in the first box will vanish. And so the distinction between the two boxes has disappeared. We can no longer tell which is **0** and which is **1**. The information is lost. So the Second Law can be stated thus: Natural processes tend to destroy information.

The Second Law explains our consciousness of the flow of time. Momentary experiences are instantly converted to memories which are constantly being stored in our brains. We remember the past but not the future because the past has left an imprint in our memories, somewhat like the sedimentary layers under the soil studied by archaeologists. As we acquire more knowledge, more information is stored in our brains. At first sight this appears to go against the Second Law, but it is not hard to see that the Second Law is not violated. The increase of information stored in the brain is accompanied by the loss of *biological information* which had been stored in the food that we digested and converted to energy. So overall, as human knowledge increases, it does so at the expense of the information that exists outside of our bodies. The storing of information in computer memories also requires energy which is ultimately obtained by the loss of information in the fuels that produce the energy. The constant supply of energy maintains the increase of information and order within our bodies, but even this process is not unending. The Second Law is also responsible for biological aging and the gradual erosion of our memories with time. The body is programmed to generate order and to decrease entropy through the intake of food and oxygen. But eventually the body will give up the fight against the tendency to greater disorder and higher entropy. Biological death is a consequence of the Second Law.

The most important thing to learn about the Second Law is that it is a *statistically based law*. Its validity rests on the very large number of atoms or molecules that make up a normal mass of matter. Maxwell — one of the pioneers of Statistical Mechanics — boldly declared: "The true logic of this world is in the calculus of probabilities."[3] Maxwell showed that Newton's Laws do not forbid heat from flowing from a cooler body to a warmer body, but he pointed out that the probability of this occurring is microscopically small.[4] So a law so fundamental as the Second Law of Thermodynamics is rooted in probability. This seemed an outright affront to the determinism that was engendered by Newton's Laws, and many philosophers and physicists resisted this statistical explanation of a basic physical law. But eventually all objections were overcome with the establishment of the atomic or molecular structure of matter.

Maxwell's dictum about the true logic of this world ultimately won the day in the triumph of quantum theory, of which he himself knew nothing. Maxwell died at the age of 48 in 1879, twenty one years before the birth of quantum theory. In quantum theory, it is probabilities that dictate the

---

[3]Lewis Campbell and William Garnett, *The Life of James Clerk Maxwell: With Selections from his Correspondence and Occasional Writings* (London: Macmillan, 1884) p. 97.

[4]In every gas the molecules are in random motion, which means not only the directions of their velocities but also the magnitudes of the velocities are random. In a hotter gas the average speed of the molecules is greater than in a cooler gas. But within each gas the molecules have a range of speeds. So the slowest molecule in a hotter gas could be much slower than the fastest molecule in a cooler gas. Maxwell devised a famous thought experiment in which a microscopic agent — called a *demon* — would allow the faster molecules from a cooler gas to pass into a warmer gas, and the slower molecules from the warmer gas to pass into the cooler gas. The result is that the average speed of the warmer gas increases, which means it gets hotter, and the cooler gas gets colder. Thus heat has been made to flow from a colder body to a hotter body without violating Newton's laws of motion. But this inference is valid only if we assume that the demon itself is not subject to the laws of physics. A material demon would get bombarded by the molecules so that it would itself execute a random motion making it totally ineffective. Statistical mechanics would prevail and heat would flow from hot to cold.

outcome of any process. The only real prediction we can make is the probability of observing one or another outcome. Maxwell's prophetic insight into microscopic phenomena made it easier for scientists to embrace the counterintuitive notions of quantum theory.

3.4 Summary

All matter is composed of indivisible particles called atoms. Atoms tend to combine with other atoms to form molecules. (Some molecules such as Helium and Argon consist of a single atom). The most important visible proof of the existence of such microscopic molecules is Brownian motion, the erratic dancing motion of pollen grains suspended in water. The laws of the combination of elements according to proportions of weight also reveal the atomic structure of matter.

The atomic structure of matter is especially discernible in the experience of heat and temperature. Heat is the energy possessed by the molecules of an object. The temperature of an object is proportional to the average kinetic energy of the molecules. Because there are a very large number of molecules in any normal quantity of a material, it is impossible to study the motions of all the molecules with precision. So we study their statistical behavior. This branch of physics is called statistical mechanics.

The study of the relationship between heat and mechanical energy is called thermodynamics. There are four important laws of thermodynamics:

(a) the zeroth law which deals with objects in thermal equilibrium with one another,

(b) the first law which states that heat can be converted to work and vice versa and there is no loss or gain of energy in this process,

(c) the second law which states that heat can flow spontaneously only from a hotter object to a cooler object, and

(d) the third law which states that it is impossible to cool down any object to absolute zero temperature in any finite number of steps.

The Second Law of thermodynamics is responsible for our subjective perception of time and memory. Time only flows forwards and never backwards. We remember the past but not the future. A fruit, an animal or a human being can only become older, never younger, as time progresses. Natural processes occur in such a way that there is a decrease in the total available energy in the universe. They also occur in such a way that information tends to get erased. We say that only those processes occur in nature that increase the total entropy of the universe.

Chapter 4

The Concept of a Field

4.1 Action at a distance

According to Newton's Third Law, when an object A exerts a force on an object B, object B simultaneously exerts an equal and opposite force on object A (Ch. 2). My weight exerts a downward force on the floor, and the floor exerts an equal upward force on my body. This upward force, called a normal force, is an example of a contact force. When I hit a ball with a tennis racket, the force applied by the ball to the racket and that applied by the racket to the ball are examples of contact forces. But the force of gravity is a different kind of force. There is no contact between the apple and the earth, and yet there is a force. There is no contact between the earth and the sun, but the force of gravity keeps the earth revolving round the sun. Gravity is an example of a non-contact force, also called *action at a distance.*

Our eyes enable us to see the sun in the sky. But how does the earth know the sun is out there, 150 million kilometers away? It feels the force of gravity. Newton did not probe into the precise mechanism by which the sun communicates its influence upon the earth. He was interested only in the effects of gravity, not how the force of gravity is communicated between objects. But this question did come up, and physicists answered it by creating the concept of a *field.* The word *field* means a region of influence between the objects that experience a mutual force. Our current understanding of the gravitational field is due to the General Theory of Relativity published by Einstein in 1915. A gravitational mass such as the sun has an influence on the space that surrounds it. This influence is communicated in all directions through the space. This influence is called the gravitational field of the sun. Any object that enters this gravitational field experiences an attraction towards the sun. The strength of the gravitational field decreases with distance. So the sun's gravitational field experienced by Mercury is much greater than that experienced by Neptune. Each planet also generates its own gravitational field. The moon is subjected to the gravitational field of the earth and is attracted to the earth. Of course, the moon also has its own gravitational field which is weaker than that of the earth because the moon is so much smaller than the earth. The actual gravitational field between the earth and the moon has a contribution from both the earth and the moon. And this is true for all pairs of objects in the universe.

So the gravitational field mediates the forces between objects. The moon cannot really "see" the earth. It "feels" the gravitational field, and this field "tells" the moon that there is an object of such a mass at such a distance away. Likewise the earth feels the gravitational field between itself and the sun, and responds accordingly.

Thus, the concept of a gravitational field eliminates the need for action at a distance. All action is contact action. An apple accelerates towards the ground because it interacts with the gravitational field between itself and the earth. It does not actually "know" there is an earth out there until it hits the ground, at which point there is a contact force between the apple and the ground.

4.2   Electricity and magnetism

Gravity was the only long range force known to humans for a long time until the discovery of magnetism and electricity. When magnetism was discovered, people were puzzled by the phenomenon. A magnetized bar of iron seemed no different from before it was magnetized — in mass, shape, color, temperature, etc. A suspended magnet aligned itself in the north-south direction. So it appeared that the magnet had tapped into some unknown power from space. Today we call it the earth's magnetic field. Every magnet has two poles, a north seeking pole and a south seeking pole, and it is impossible to isolate a magnetic pole. If we were to divide a magnet in half, we would get two smaller magnets, each with a north and a south pole. The difference between the magnetic and the gravitational fields is that whereas the gravitational field always produces attraction between two objects, a magnetic field produces repulsion between two like poles and attraction between unlike poles. Similar to the gravitational field, the magnetic field due to a magnetic pole becomes weaker with the distance from the pole.

A magnetic field is often represented in books by lines called magnetic lines of flux. Lines can be imagined as flowing out of a north seeking pole and ending in a south seeking pole. The direction of the force experienced by a magnetic pole at any point in a magnetic field is given by the direction of the flux line at that point. A north pole would tend to move in one direction along the line, and a south pole would tend to move in the opposite direction.

The source of magnetism had to remain a mystery until the discovery of electricity. It was surely one of the most thrilling moments in the history of science when they found that a magnet was affected by the flow of electric current in a nearby wire.

Electric current is the flow of electric charges. And like magnetic poles, electric charges come in two opposite varieties, which we call positive and negative. Similar to magnets, unlike charges attract and like charges repel. But in contrast to magnets, it is possible to isolate charges. Every charge generates its own electric field. So a positive charge will sense the presence of another charge in the vicinity by picking up the electric field produced by the other charge. This field will tell the positive charge if the other charge is positive or negative. The strength of the field diminishes with distance, just like magnetic and gravitational fields.

Electric charge is measured in coulombs (C). The force between two charges q_1 and q_2 at a distance r from each other has a magnitude equal to

$$\frac{q_1 q_2}{4\pi\epsilon_0 r^2} \text{ newtons (N)} \tag{4.1}$$

where ϵ_0 is a constant called the permittivity of free space $= 8.85 \times 10^{-12}$ in SI units. The direction of the force is along the line between the charges, and, as stated above, like charges repel and unlike charges attract.

The smallest charge that can be found in nature is the charge of the electron which has the value of -1.6×10^{-19} C. The proton has an equal and opposite (and therefore positive) charge.

A charge has a region of influence or electric field that extends outwards in all directions. As the distance increases, the field decreases according to the square of the distance as shown in Eq. (4.1).

Exercise 4.1. What is the magnitude of the force between a proton and an electron at a distance of 1.4×10^{-8} m from each other?

In general, an electric field is the result of the presence of several charges, some positive, some negative, some stationary, some moving. The strength or *intensity* of an electric field varies from point to point in both magnitude and direction and is represented by the vector **E**.[1] Suppose a stationary charge q is placed in an electric field of intensity **E**. It would experience a force given by $q\mathbf{E}$. (**E** itself arises from the presence of other charges, positive or negative, stationary or moving.) So we have the equation for the force on a charge

$$\mathbf{F} = q\mathbf{E}$$

An electric force field is represented by field lines or lines of force. Lines are drawn flowing out of a positive charge and ending in a negative charge. The direction of the force experienced by a positive charge at any point in an electric field is given by the direction of the field line at that point. A positive charge would tend to move in one direction along the line, and a negative charge would tend to move in the opposite direction.

4.3 Electromagnetism

Magnetism and electricity are closely related phenomena. A changing electric field produces a magnetic field and a changing magnetic field produces an electric field. If we have a wire carrying current, then a magnetic field is created that circles round the wire as shown in Fig. 4.1. We notice that in this case the magnetic line of flux is a complete circle — it does not originate in a north pole or terminate in a south pole.

If a current moves in a circle, such as a loop, we can show that the magnetic lines of force will curl round the wire of the loop in such a way that the lines come out of one face of the loop and go in through the other face. So the loop acts like a magnet, with a north pole on one face and a south pole on the other face.

In a long coil of wire — called a *solenoid* — carrying current, one end is a north pole and the other end a south pole. To figure out which is which, look at the direction of the current on each end face. If the current is counterclockwise, that face is North. If the current is clockwise, that face is South.

---

[1] A quantity with magnitude and direction is called a *vector*. Textbooks usually indicate a vector by a letter in bold face type, say **V**. The magnitude of this vector is usually expressed by the same letter in ordinary type, V.

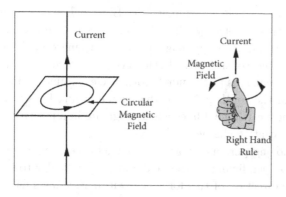

Fig. 4.1. A wire carrying current produces a circular magnetic field.

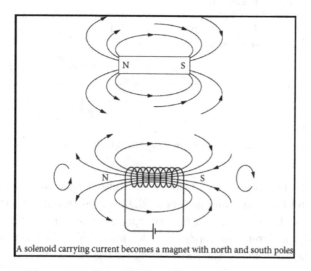

A solenoid carrying current becomes a magnet with north and south poles

Magnetism in iron and other magnetic substances is caused by tiny circular currents. These currents are produced by spinning and orbiting electrons. Each atom is therefore a tiny magnet. In a regular rod of iron, these tiny atomic magnets are arranged in random haphazard directions, so that the magnetism of the individual atoms all cancel each other. But when the rod is magnetized, all the atoms align themselves like soldiers in formation. Then one end of the rod becomes a north pole and the other end a south pole. In a real iron bar, the atomic magnets tend to be aligned with their immediate neighbors to form *domains*. Within each domain all the atomic magnets are parallel. Neighboring domains are randomly oriented, so that the bar as a whole is not magnetic. But when it is magnetized, the domains all align themselves so that the bar becomes a magnet.[2]

The flux lines due to a bar magnet do not originate in the north pole or terminate in the south pole. Each line is a closed loop that travels through the bar, emerges at the north pole, circles around and reenters the magnet through the south pole. So the magnetic flux lines are all closed loops, unlike the electric field lines which originate in a positive charge and terminate in a negative charge.

A charge moving perpendicular to a magnetic field experiences a force in a direction perpendicular to the field and perpendicular to its direction of motion. The direction of this force is given by Fleming's left-hand rule.

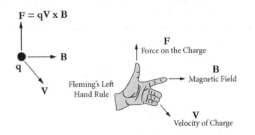

Fig. 4.2.   A charge moving in a magnetic field experiences a force.

Fleming's left-hand rule: Extend the thumb upwards, the forefinger forwards, and the remaining fingers at right angles to both the thumb and the forefinger. If the remaining fingers represent the velocity of the positive charge and the forefinger represents the magnetic field then the thumb indicates the direction of the force acting on the charge.[3]

We write the expression for the force acting on a charge moving in a magnetic field as

$$\mathbf{F} = q\mathbf{v} \times \mathbf{B} \tag{4.2}$$

where the multiplication sign is called the cross product. This vector symbol means that the magnitude of the force is given by $F = qvB\sin\theta$ where θ

---

[2]A magnetized bar of iron has greater order and therefore lower entropy than the bar before it was magnetized. After a very long time, the bar will lose its magnetism as it loses its order and thereby gains entropy in keeping with the Second Law of Thermodynamics.

[3]The direction of the force is opposite for a negative charge moving with the same velocity in the same magnetic field.

is the angle between the velocity of the charge and the magnetic field, and the direction of the force is at right angles to both the velocity and the magnetic field.[4] (See the footnote in Ch. 2 for the definition of $\sin\theta$.)

Exercise 4.2. A particle having charge 0.3 C is traveling at a speed of 2.5×10^3 m/s and enters a magnetic field that is perpendicular to the direction of motion of the particle.

(a) If the strength of the magnetic field is 4.0 T (tesla = SI unit of magnetic field strength) find the magnitude of the force acting on the particle.

(b) If the mass of the particle is 4.0 kg, find the magnitude of the acceleration imparted by the magnetic field to the charged particle. (Use Newton's Second Law.)

Every charge moving in a magnetic field experiences a force. An electric current is simply a stream of charges in motion. A conductor carrying current will therefore experience a force in the presence of a magnetic field. This is the principle used in electric motors. The essential part of an electric motor consists of an electric current passing through a rotatable conductor placed in a magnetic field. The motion of the charges along the conductor will cause the charges, and hence the entire conductor, to experience a force. By means of a suitable arrangement a continuous current that passed through the conductor in a magnetic field causes the conductor to rotate continuously.

So we can generate a force on a charge by moving it through a magnetic field. But it turns out that it is not necessary for the charge to be in motion at all. What actually matters is the *relative motion* between the charge and the magnetic field. If a charge is stationary, and a magnet is brought towards it, the charge will experience a force. Bringing the magnet will cause a magnetic field to move relative to the charge. Now, we already know that stationary charges are affected only by an electric field and not by a magnetic field. So if a stationary charge is affected by a changing magnetic field, this can only mean that *a changing magnetic field produces an electric field!* If we bring a bar magnet close to a loop of wire charges will flow around the loop and thus there will be a current in the loop. This

---

[4]A line drawn at right angles to the page of this book can be going into the page or coming out of the page. Fleming's Left Hand Rule eliminates this ambiguity from the cross product.

shows that a changing magnetic field produces a circulating electric field, which is different from the electric field produced by a stationary charge. This has useful practical applications. We can use a changing magnetic field to produce a current in a loop of wire. This is the principle used in generators that produce electric current.

We can now sum up all the laws we have learned so far in the following four rules:

1. A moving electric charge generates a magnetic field as seen in Fig. 4.1. A moving charge carries its electric field with it. When a charge is approaching us we experience an increase of electric field intensity. A charge receding from us will generate a decrease in electric field intensity. In either case a magnetic field is generated. *A varying electric field produces a magnetic field.*

2. Electric charges can be isolated. A positive charge acts as a source of an electric field. Field lines flow outwards from a positive charge and flow into a negative charge.

3. A stationary charge is not affected by a magnetic field. But if a magnet is moved close to a stationary charge, this charge will feel a force, because the changing magnetic field will produce an electric field that affects the charge. *A varying magnetic field produces an electric field.* This electric field produced by a varying magnetic field does not flow out of a positive charge or into a negative charge but has the form of a closed loop.

4. It is impossible to isolate a magnetic pole. Magnetic flux lines are always closed loops.

The above four statements were expressed as mathematical equations by Maxwell. They are called Maxwell's Equations, and they provide the mathematical foundation for the classical theory of electricity and magnetism.

4.4   Electromagnetic waves

J. C. Maxwell was the greatest physicist of the nineteenth century. His contributions to statistical mechanics are little short of monumental. But his greatest achievement lies in his unification of electricity and magnetism. Maxwell collected all the discoveries that physicists had made concerning electricity and magnetism and united all the known laws into a single mathematical theory. Maxwell showed that electricity and magnetism are different manifestations of the same phenomenon. Electric charges can be isolated, but there is no such thing as an isolated magnetic pole. Moving charges constitute electric current, and electric current produces magnetism. Thus magnetism is an effect that arises from the motion of electric charges. Two positive charges A and B of magnitudes q_1 and q_2 coulombs respectively at a distance r meters from each other experience a force of repulsion F newtons given by the expression

$$F = \frac{q_1 q_2}{4\pi\epsilon_0 r^2} \tag{4.3}$$

Now suppose the charge B were to be displaced by a small distance so that the relative displacement between the two charges is now changed. Clearly the force between them would change in direction or magnitude or both.

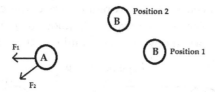

So if the initial force is called F_1 and the final force F_2, these forces would have different directions and different magnitudes. Here is a question: would charge A respond to the displacement of charge B instantaneously, or is there a time lag between the occurrence of the displacement of B and this displacement being "noticed" by A? The answer given by Maxwell's electromagnetic theory is that the electric force is not communicated instantaneously from one charge to another, but that it takes time to travel between the charges. So, if one charge were displaced slightly, it would take some time — however short — before the second charge experiences a change of force. These electrical influences travel at a great but finite speed, and Maxwell showed that this speed is also the speed of light.

Consider a stationary electric charge at a point P. This charge has an electric field that extends in all directions. If this charge is given a sudden jerk, it would generate a magnetic field because of its motion. This magnetic field would also move outwards. Now consider a point Q a short distance from the charge. Originally there was only an electric field at this point due to the charge. Now there is also a magnetic field due to the motion of the charge. Thus the magnetic field at this point has changed, having gone from zero to some value. This changing magnetic field in turn generates its own electric field which in addition to the original electric field results in a changed electric field at the point Q. This change of electric field will in turn generate a magnetic field, and so a cycle is set up. Both the changing electric and magnetic fields will travel outwards from the charge in all directions in an electromagnetic dance. We call this an electromagnetic wave.

Of course, this wave will die out. But if a constant supply of energy is given to the charge so that it is made to continually vibrate up and down along a conductor, it will constantly emit electromagnetic waves that radiate away from the conductor. This is what takes place in a broadcasting antenna of a radio or television station or even your cell phone.

An important discovery of the nineteenth century was that electromagnetic waves travel at the speed of light, which is 3×10^8 m/s. It was a small but major step to infer that light itself is a form of electromagnetic wave. The difference between light and a radio wave is the difference in wavelength. Light has a very short wavelength, about 500 nm (nanometers) = 5×10^{-7} m whereas a radio wavelength can be of the order of 500 m.

Shape of a Wave

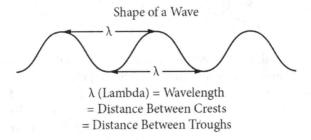

λ (Lambda) = Wavelength
= Distance Between Crests
= Distance Between Troughs

4.5   Finite speed of fields

So any wave in an electromagnetic field travels at the speed of 3×10^8 m/s. Maxwell's theory provided inspiration for Einstein to develop his Theory of Relativity according to which nothing can travel faster than 3×10^8 m/s. Einstein also showed that light and other electromagnetic waves can never be found traveling slower than this speed. This will be the subject of a later chapter.

Einstein showed that gravitational fields also cannot travel faster than 3×10^8 m/s. It is the gravity of the sun that keeps the earth moving in a circular orbit. If this gravity were to disappear suddenly the earth would fly off along a tangent. It takes about 8 minutes for light to reach the earth from the sun. If the sun were to disappear suddenly, its effect would not be felt on earth instantly, but there would be a lag of 8 minutes between the disappearance of the sun and the earth shooting off into outer space.

We saw above that a vibrating electric charge emits electromagnetic waves. Similarly, a vibrating mass emits gravitational waves. But because the force of gravity is so much weaker than the electromagnetic force, it is difficult to detect gravity waves. If two very massive objects collide with each other that would also send a gravitational wave pulse outwards through space. Such gravitational waves have been detected.

Because all fields communicate energy at finite speeds we can think of this energy as traveling through the field from the source to the detector. In the next chapter we will investigate the shape of the energy communicated in an electromagnetic field. And that will bring us to the genesis of quantum theory.

4.6 Summary

When I push against a wall, the wall exerts an equal and opposite force on me. These forces are examples of contact forces. If I throw a ball into the air it will come down because of the force of gravity. Gravity is an example of a non-contact force or action at a distance. Other examples of non-contact forces are electrical and magnetic forces.

Non-contact forces are mediated by fields — gravitational fields and electromagnetic fields. When an object is suddenly jerked, it produces an effect in the field immediately surrounding it. This effect travels as a wave outwards from the object through the field. So, an oscillating electric charge produces an electromagnetic wave. A massive object that oscillates rapidly generates a gravitational wave.

Electricity and magnetism are related to each other. An electric field is created by an electric charge. When a second charge is brought in the field of the first, the two charges experience forces of attraction if they are unlike charges and repulsion if they are like charges.

A moving electric charge causes the electric field surrounding it to change. A changing electric field produces a magnetic field. And a changing magnetic field produces an electric field. So a vibrating electric charge will generate a stream of oscillating electric and magnetic fields that move outwards from the charge. This is called an electromagnetic wave. All electromagnetic waves travel at the same speed, which is 3×10^8 m/s.

Electromagnetic waves differ in their wavelength. Light is a form of electromagnetic wave, having wavelengths much shorter than those of radio waves.

Chapter 5

The Ultraviolet Catastrophe

5.1   A black body

Maxwell showed that the electromagnetic wave generated by a moving charge travels through space at the speed of light which has the value of $c = 3 \times 10^8$ m/s. This led to the inevitable conclusion that light itself is nothing but a form of electromagnetic radiation. The difference between light and the electromagnetic waves that we pick up with our cell phones, TV sets, radios and cell phones is that these latter have longer wavelengths than light waves. Light waves also have a range of wavelength. Light having wavelength between 620 and 750 nanometers (1 nanometer = 10^{-9} m) appears red and at the other end of the spectrum light with wavelength between 380 and 420 nanometers appears violet to the average human eye.

It is not just light that has different colors. When light falls on objects we see them as having color. Leaves are mostly green because they have chlorophyll, a pigment that absorbs light of most wavelengths but does not absorb the wavelengths corresponding to green light, which explains why leaves appear green in normal sunlight. In general, an object has a particular color because it does not absorb light of that color but instead scatters light having wavelength corresponding to that color.

A black object such as charcoal or soot has this property that it absorbs virtually all colors. Since it does not scatter any wavelength it does not have any color, but appears black. An actual black object is not a perfect absorber of light, but in this chapter we will consider an ideal black surface that absorbs every wavelength of electromagnetic radiation. Such a black surface — called a *black body* — would also emit electromagnetic radiation of all wavelengths.

5.2   Black body cavity

Let us now imagine that we have a cavity enclosed by walls that are perfectly black. Suppose we were to make a small opening in this cavity and introduce some electromagnetic radiation and close up the opening immediately. What would happen to this radiation?

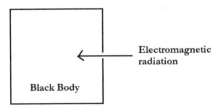

A cavity in the shape of a hollow cube

When this radiation hits the walls of the cavity it would be entirely absorbed, because the black body is a perfect absorber. The wall that absorbed the radiation would also emit radiation, though not necessarily at the same frequency. The walls continue to absorb and emit radiation. Thus there is a constant exchange of energy between the cavity and the walls. This process is analogous to the case of the gas within a closed container, but there is one significant difference. Whereas in a gas the molecules exchange energy with the walls and *also with each other*, the radiation in the cavity can exchange energy only with the walls. Electromagnetic waves pass right through each other without colliding.

Light waves pass
through each other

At equilibrium the walls absorb and emit radiation and this radiation flows from one wall to another. The emitted radiation need not have the same wavelength as the absorbed radiation. A black wall is free to emit radiation of any frequency. Moreover, the walls are free to emit radiation in any direction, not just the direction of the absorbed radiation. However, by conservation of energy, the walls cannot emit more energy than they absorb. Since the walls can emit radiation at all possible frequencies, they will continue to emit radiation at both higher and lower frequencies than the original radiation that was introduced into the cavity. The laws of statistical mechanics demand that the energy be divided equally among all the various degrees of freedom within the container. And the different frequencies correspond to different degrees of freedom for electromagnetic waves.

5.3   Standing waves and the catastrophe

Now, the laws of wave motion require that a closed container cannot generate radiation having wavelength longer than the dimensions of the container. The longest possible wavelength is twice the length of the container, cf. the topmost wave in the figure on the left. But there is no restriction on the shortness of the lengths of the waves.

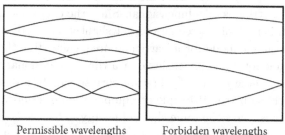

Permissible wavelengths   Forbidden wavelengths

When there is equilibrium, there is a stable distribution of frequencies of radiation. When electromagnetic radiation is in a stable state within a cavity, only certain wavelengths are permitted. For the radiation to be in a stable state, it must exist as a *standing wave* between the two walls, much like the standing waves in a string that has been clamped at its two ends but free to vibrate in between.

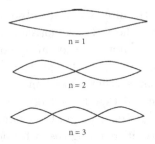

The figure shows a string of length L vibrating in its three lowest modes: the fundamental or first harmonic, the second harmonic and the third harmonic. The wavelength is different for each mode. For the first mode the wavelength is $2L$, for the second it is L (or $2L/2$), and for the third it is $2L/3$. Proceeding this way, the wavelength of the nth mode would be $2L/n$. As the wavelengths decrease, the frequencies increase, since the frequency ν is related to the wavelength λ and the speed of the wave c by the formula $c = \nu\lambda$. (The Greek letter *nu* written as ν is commonly used to denote frequency. It is very different from the symbol v used for velocity.)

A vibrating string is one-dimensional. Since the cavity we are considering is three-dimensional, the equations get slightly modified, so that there can be more than 1 mode for a particular frequency. But the end result will be the same for a one-dimensional or a three-dimensional system.

Each one of these modes of vibration corresponds to a distinct degree of freedom, and so by the Principle of Equipartition of Energy each one of these modes must have the same share of energy.

But how many such modes are there? We saw above that there is an upper limit to the wavelength but there is no lower limit. The wavelengths of the modes can become as small as possible — all the way to zero. And correspondingly there is no upper limit to the frequency. It can become as great as possible — all the way to infinity.

If there is no upper limit to the frequency there is clearly no upper limit to the number of modes that can be created inside the cavity. And the available energy has to be distributed evenly among all the modes which now become unending.

What this means physically is that the walls of the cavity will continue to generate waves of shorter and shorter wavelength without ever stopping. So if the original radiation was microwave radiation, then as time progresses the cavity would contain not only microwaves but also visible radiation with all frequencies between red and violet, and it would not stop there but the cavity would relentlessly continue to generate waves of shorter and shorter length. And as this continues, the average wavelength of the radiation in the cavity would become shorter and shorter and the corresponding frequency becomes higher and higher without any limit. Thus the cavity would be sliding into an unfathomable abyss of perpetually increasing frequency going way beyond ultraviolet, X rays and γ rays.

Thus, the cavity would never reach equilibrium. Historically this was called the *ultraviolet catastrophe*, where the word *ultraviolet* simply meant frequencies higher than those corresponding to violet light, including X rays and γ rays. This was the theoretical prediction.

However, experiments on real cavities which were excellent approximations to the black body cavity showed no such tendency. Indeed, equilibrium was reached in a very short time, and there was no tendency on the part of the radiation to move to shorter and shorter wavelengths. So the catastrophe was fictitious rather than real. But then, there seemed to be something seriously missing in the physics of the late nineteenth century.

5.4 Escape from the ultraviolet catastrophe

It turned out that the only way out of the ultraviolet catastrophe was to make a most extraordinary assumption, that the walls of the cavity could only absorb or emit radiation with energy equal to the frequency multiplied by a constant factor. This was a revolutionary assumption, and made no sense according to the physics of the day. The only thing that could be said in its favor is that *it worked!* This assumption was first made by Planck, who obtained an expression for the distribution of wavelengths inside a cavity that agreed perfectly with experiment. How did this assumption avert the ultraviolet catastrophe?

The assumption stated that the energy of the electromagnetic radiation that could be absorbed or emitted by the walls is given by $\epsilon = h\nu$ where ν is the frequency of the radiation. Thus, electromagnetic radiation is absorbed or emitted in discrete units. And each such unit was called by the Latin word for unit *quantum*. The constant of proportionality h has since been called Planck's Constant.

Clearly, not all quanta have equal energy. The energy of a quantum is proportional to the frequency of the radiation. The higher the frequency, the greater the energy. So the frequencies corresponding to the first three modes will be the following:

First harmonic: $\nu_1 = \frac{c}{2L}$

Energy of 1 quantum of the first harmonic $= h\left(\frac{c}{2L}\right)$. Let us call this E.

Second harmonic: $\nu_2 = 2\left(\frac{c}{2L}\right)$

Energy of 1 quantum of the second harmonic $= 2h\left(\frac{c}{2L}\right) = 2E$

Third harmonic: $\nu_3 = 3\left(\frac{c}{2L}\right)$

Energy of 1 quantum of the third harmonic $= 3h\left(\frac{c}{2L}\right) = 3E$

Exercise 5.1. A cubical cavity has length 0.1 m. Show that the smallest frequency of a standing wave within this cavity is 1.5×10^9 Hz. Speed of electromagnetic waves $= 3 \times 10^8$ m/s.

As the number of the mode increases, the energy of one quantum of that mode increases proportionately. But since the total amount of available energy inside the cavity is finite, eventually we will reach a mode of sufficiently high number that cannot even have a single quantum. The lower modes can have several quanta each, but the number of possible quanta will decrease at higher energies and will stop altogether once a ceiling has been reached.

Imagine a jungle party hosted by some little rodents at which (herbivorous) animals of different sizes are invited, but there is a limited quantity of food. Each animal is promised at least one mouthful, with the specification that they get only entire mouthfuls, not partial mouthfuls. But a mouthful for a hippopotamus is much bigger than a mouthful for a rabbit. If the total amount of food that was available is less than a hippo's mouthful, then this poor animal would have to be turned away, and of course there would be no food for a larger animal like an elephant. This means that all animals with mouths as big as or bigger than a hippo's cannot expect to get any food at the party.

Translating this to the black body cavity, all the frequencies higher than a particular number cannot expect to get any energy from the cavity, because they can only take energy in multiples of quanta, where each quantum has the value $h\nu$. If the total amount of energy available at the beginning is E, then there is a frequency ν_{max} such that $h\nu_{max} = E$ so that no radiation having frequency greater than ν_{max} can be generated inside the cavity. There is an upper limit to the frequencies. Thus the ultraviolet catastrophe is averted. Quantum theory provides an escape from this perpetual spiraling towards unending higher frequencies.

Exercise 5.2. Show that ν_{max} within a cavity into which 1.5×10^{-12} joules of energy are introduced must be 2.3×10^{21} Hz. $h = 6.6 \times 10^{-34}$ m^{2}kg/s.

5.5 A small beginning

And so quantum theory was born. It was a relatively quiet birth, without much fanfare. Planck's resolution of the problem of the ultraviolet catastrophe was a mathematical solution. The mathematical formula that expressed the distribution of the frequencies inside a blackbody could be correctly derived only by assuming that energy was absorbed and emitted in quanta proportional to the frequency. This is called the quantum hypothesis. It seemed to be a small correction to an otherwise reliable system of physics.

Though the quantum hypothesis had shown that classical physics could not explain the energy distribution in black body radiation, that did not seem a big deal. It merely appeared that there was a small crack in the mighty edifice of classical physics.

But the crack grew wider, and the structure of physics that took centuries to build finally collapsed. Physics would never again be the same. In the following chapters we will see the steps by which a new understanding of reality gradually developed following the initial step taken by Planck's quantum hypothesis.

5.6   Summary

The law of equipartition of energy of statistical mechanics states that the average kinetic energy of a system is distributed evenly among all the degrees of freedom possessed by the system. A Helium gas has monatomic molecules that have three degrees of motion, because they are capable of moving in all three dimensions. The average kinetic energy of the molecules is divided equally among the components of the motions of the molecules along three mutually perpendicular directions. This is an example of equipartition.

A surface that absorbs every wavelength of electromagnetic radiation also emits every wavelength of electromagnetic radiation. Such a surface is called a black body. A cavity whose inside walls are coated with a black body substance is called a black body cavity. If some radiation is introduced inside this cavity, then after repeated absorption and emission by the walls, all possible frequencies are generated, and the available energy is shared equally among all the possible frequencies. This leads to the ultraviolet catastrophe, since there is no upper limit to the frequency, and the average frequency of the radiation will continually increase without stopping.

The ultraviolet catastrophe is averted by assuming that energy can be absorbed or emitted only in multiples of the frequency, expressed by the formula $E = h\nu$ where ν is the frequency of the radiation, E is the energy of a unit (called quantum) of radiation, and h is called Planck's Constant.

Chapter 6

Absorption and Emission of Radiation

6.1 Photelectric effect

Electrical current is caused by electrons traveling about in a conductor. Ordinarily an electron does not jump out of a conductor. There are cohesive forces that prevent electrons from separating themselves from the conducting object within which they move. In order to entice an electron to escape from the surface of the conductor we need to offer a strong force of attraction to the electron. This can be done by bringing an opposite — that is, positive — charge close to the conductor. If this charge is sufficiently high, then it will create a high electric field in its vicinity. The effect of this electric field on the electrons in the conductor would be to draw the electrons towards the positive charge. Some of these electrons would jump from the surface of the conductor to the positive charge that has been brought close. This flow of electric charges through the air from one conductor to another is visibly seen as a spark discharge. Such discharges require very strong electric fields. Lightning is an example of a spark discharge.

But there is another phenomenon in which electrons can be released from a conductor without the use of a very strong electric field. When light waves fall upon certain metallic conductors electrons are released from the surface of the conductor. This is called the photoelectric effect. This effect was discovered when certain metallic conductors which were bombarded with electromagnetic radiation loosened up electrons which were attracted by a relatively small electric field outside of the conductor.

The following is a simplified description of an apparatus that illustrates the photoelectric effect. Two conducting plates are kept a small distance apart in a vacuum. One plate is connected to the negative terminal of a battery or some other DC (direct current) power supply and the other plate is connected to the positive terminal. Thus there is an electric field set up between the positive and the negative plates which attracts the electrons on the negatively charged metal plate. They cannot leave the surface of the plate because of the cohesive forces that bind them to the body of the conducting plate. However, if electromagnetic radiation such as light is shone on the plate, this radiation can communicate energy to the electrons enabling them to overcome the cohesive forces and leap through the vacuum to the positive plate. Once inside the positive plate they are attracted to the positive terminal of the power supply and flow through the measuring device called the ammeter. This ammeter records the amount of electric current — which is proportional to the rate at which electrons flow through the device.

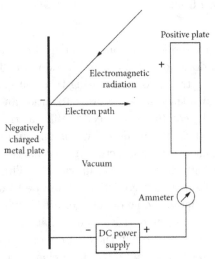

The results of the experiments can be summed up as follows:

1. When the negatively charged plate was bombarded with radiation of low frequency, there was no discharge of electrons, no matter how great was the intensity of the radiation. Thus even a large amount of energy given to the conductor did not serve to release any electrons from the conductor.

2. Then the frequency of the radiation was increased steadily. At a particular frequency, the electrons were emitted. Keeping the frequency of the radiation constant, the intensity of radiation was increased. As the intensity of radiation increased, more electrons were released. The current measured by the ammeter increased in proportion to the intensity of the radiation.

3. The frequency of the radiation was now raised, without changing the intensity. Now it was observed that the electrons were leaving the plate with greater speed, or greater kinetic energy. But the number of electrons did not increase.

As a particular example, the metal potassium did not emit any electrons when red light (low frequency) fell upon it. When green light (higher frequency) was shone on the metal it began to emit electrons. The energy of the electrons increased as the frequency of the light was further increased to blue and violet.

So these results lead to the following inferences.

1. A minimum frequency of radiation is necessary to release the electrons, i.e. to help them overcome the cohesive force that binds them to the conductor.

2. Once this minimum frequency was reached, the number of electrons released is proportional to the intensity of the radiation.

3. As the frequency of the radiation is increased, the energy of the electrons also increases.

6.2   Einstein's explanation

Albert Einstein provided a simple explanation for this photoelectric effect. He suggested that electromagnetic energy travels in the form of packets of energy, with each packet having energy proportional to the frequency of the wave, and on the basis of this suggestion he was able to explain the photoelectric effect.

Planck had suggested that radiation was emitted or absorbed in units or quanta with energy $\epsilon = h\nu$ where ν is the frequency of the radiation. Einstein's postulate carries Planck's hypothesis to the next logical step — that radiation is not only absorbed or emitted in quanta, but radiation *exists* only as quanta of energy $h\nu$.

Einstein's explanation for the photoelectric effect can be summed up as follows:

Each quantum of electromagnetic radiation has energy $h\nu$. An electron in the metallic conductor receives this energy from the bombarding radiation. If this energy is low, the electron cannot escape from the metal surface. As the frequency ν is steadily raised, a point is reached when the energy of each quantum of radiation is exactly equal to the energy needed for the electron to burst free of the cohesive force. If the intensity of the radiation is increased, then more quanta are present in the radiation. Each such quantum gives its energy to an electron, and releases the electron. Thus the number of released electrons — which is measured as the current by the ammeter — is proportional to the intensity of the radiation, since the number of emitted electrons is equal to the number of quanta in the radiation. As the frequency of the beam is increased, the energy of each quantum increases according to $\epsilon = h\nu$. This means that each electromagnetic quantum is able to give more energy to each electron.

Exercise 6.1. Light of frequency 7.0×10^{14} Hz was shone on a metal surface. Electrons of kinetic energy up to 1.30×10^{-19} J were emitted. The frequency of the light was raised to 8.0×10^{14} Hz. This time electrons of kinetic energy up to 1.96×10^{-19} J were emitted. The relationship between the energy of the light ($\epsilon = h\nu$) and the maximum kinetic energy T of an emitted electron is the following:

$$\epsilon = T + \Phi$$

where Φ is a constant number equal to the work done by the electron in escaping from the metal surface. Use the data given in this exercise to calculate the value of h.

6.3 Momentum of electromagnetic radiation

When electromagnetic energy travels from place to place at the speed c (which is the speed of light $= 3 \times 10^8$ m/s) it also carries momentum with it. According to classical electromagnetic theory this momentum is equal to the energy divided by c, the speed of the radiation.

Applying this to the quantum, we obtain an expression for the momentum of a quantum of radiation:

$$p = \frac{h\nu}{c} = \frac{h}{\lambda} \tag{6.1}$$

where λ is the wavelength of the radiation (equal to c/ν).

Thus a quantum of electromagnetic energy has energy $h\nu$ and momentum h/λ.

This suggests that a quantum of radiation behaves somewhat like a particle in that it has a definite energy and a definite momentum. So the quantum theory suggests that a beam of electromagnetic energy of a definite wavelength (or frequency) is communicated as tiny particles all having the same energy $h\nu$ and the same magnitude of momentum h/λ.

These particles are called *photons*. A photon is a particle of light or other electromagnetic radiation. A photon is characterized by the wavelength or the frequency of the radiation. All photons travel at the speed c.

Does a photon have mass? If it has energy and momentum, it makes sense to say that a photon has mass. Let the mass of a photon of frequency ν be m. The relation between momentum p and mass m for an ordinary particle such as an electron is $p = mv$ where v is the speed of the particle. So for light we would have: $p = mc$. Since $p = h/\lambda$, we can write $mc = h/\lambda$ and so $mc\lambda = h$. Multiplying both sides by the frequency ν, we obtain $h\nu = mc^2$. Since $h\nu$ is the energy of the photon, which we can write as ϵ, we obtain the important equation

$$\epsilon = mc^2 \tag{6.2}$$

This is Einstein's famous energy mass equation, which he had obtained using a different method, and which applies to all sorts of matter, not just photons. The method Einstein used is called the Special Theory of Relativity, which Einstein published the same year that he published the theory of the photoelectric effect. We will discuss this theory in Ch. 8.

Exercise 6.2. Find the momentum of a light photon having wavelength 235 nm (1 nm = 10^{-9} m). Next find the mass of the photon. Take the speed of light $c = 3.0 \times 10^8$ m/s.

6.4 Compton effect

So a photon has energy, mass and momentum. If it is absorbed by an electron, the electron would acquire the energy of the photon. What about the momentum of the photon? Does the electron also acquire the momentum of the photon it absorbs? The answer to this question has been provided in the affirmative. The evidence is a process called the Compton Effect. In this effect a photon collides with a stationary (or slow moving) electron, and gives the electron a certain amount of momentum. The electron then emits a photon of lower frequency and therefore lower energy. The difference in energy between the absorbed and emitted photons is equal to the increase in energy of the electron. This increase of energy is manifested as kinetic energy, due to the motion of the electron. Also, the change in momentum of the electron equals the difference in momentum between that of the emitted photon and that of the absorbed photon. Since momentum is a vector, a change of momentum includes change of direction as well as change of magnitude.

Applying the conservation of momentum and energy to the case of an absorption followed by emission of a photon by an electron that was initially at rest, if m_0 is the mass of the electron, λ_i the wavelength of the absorbed photon, λ_f the wavelength of the emitted photon, and θ the angle between the paths of the initial and the final photons, a detailed calculation yields the relation

$$\lambda_f - \lambda_i = \Delta\lambda = \frac{h}{m_0 c}(1 - \cos\theta) \tag{6.3}$$

This process is called the Compton Effect. The process is usually described not as an absorption followed by an emission, but as a scattering. The absorbed photon is called the incident photon and the subsequently emitted photon is called the scattered photon. The Compton Effect vividly illustrates the particle nature of the photon, without detracting from the fact that the photon is indeed a quantum of the electromagnetic field.

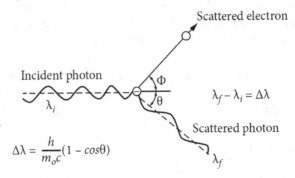

Compton Effect

Exercise 6.3. If a photon is incident on a stationary electron and is scattered through 90^0, the change in wavelength of the photon $\Delta\lambda$ is called the Compton Wavelength of the electron. Using the formula of Eq. (6.3) find the Compton Wavelength of an electron using the following data: Planck's Constant = 6.6×10^{-34} m^2 kg/s, electron mass = 9.1×10^{-31} kg and speed of light = 3.0×10^8 m/s. Check your result with some reliable online source.

6.5 Finite time of interaction

Quantum theory was born when it was realized that electromagnetic energy can be absorbed or emitted only in units called quanta, and the energy of each quantum is given by $\epsilon = h\nu$ where ν is the frequency of the radiation.

The major step — which we may rightly call a quantum leap — made by physics at this stage of history was the realization that electromagnetic energy cannot be absorbed or emitted continuously. This was such a new concept that it was necessary to come up with a whole new way of describing the process of absorption or emission of electromagnetic energy.

In classical physics there are numerous processes where energy is exchanged between two objects. Consider a charged body of mass m that is accelerated by absorbing electromagnetic radiation. The charged body receives electromagnetic energy from the field and its kinetic energy increases from T_i to T_f in the amount

$$T_f - T_i = \frac{1}{2}mv_f^2 - \frac{1}{2}mv_i^2 \tag{6.4}$$

In classical physics $T_f - T_i$ can have any of a continuous range of values from 0 to some maximum value.

Exercise 6.4. A ball of mass 0.12 kg is thrown upwards at 10.0 m/s. After it has risen to a certain height its speed decreases to 2.0 m/s. How much kinetic energy did the ball lose during this interval of time?

But in quantum physics such a continuous exchange of electromagnetic energy is impossible. Energy must be received or given away in multiples of $h\nu$ where ν is the frequency of the electromagnetic radiation.

How does this change the way we understand interactions between charges and fields? Specifically, let us talk about the duration of an interaction. Let us consider a charged body such as an electron interacting with an electromagnetic field. Let us say that this interaction took place over a time interval T. And let us say that during the time interval T a certain amount of energy E was exchanged. So the rate of energy exchange equals E/T. Classical physics claimed that in the course of half this time interval, i.e. in the time $\frac{T}{2}$, the amount of energy that was exchanged was $\frac{E}{2}$ (if we assume the rate of energy exchange was constant). By arbitrarily decreasing the time of interaction we could correspondingly decrease the amount of energy that was exchanged according to classical electromagnetism.

But Planck has shown that such a continuous exchange of energy is not possible. Thus it is impossible to decrease the time of interaction to arbitrarily small values.

Consider the standing waves within the black body cavity. The smallest mode — corresponding to the lowest frequency and greatest wavelength — has a wavelength equal to twice the length of the cavity. This mode has an energy equal to $h\nu$ where ν is the frequency of the radiation.

When the wall absorbs energy from the cavity, it absorbs the entire quantum, not a part of it. Now, since electromagnetic radiation travels at a finite speed c, the time taken for the wall to absorb the quantum of radiation is related to the time taken for the quantum of radiation to travel the length of the cavity. Let us call this time Δt. So $\Delta t = \frac{L}{c}$. Now $L = \frac{\lambda}{2}$. So $\Delta t = \frac{\lambda}{2c} = \frac{1}{2\nu}$. This yields the relation $\nu \Delta t = \frac{1}{2}$.

Multiplying both sides by Planck's constant h we obtain

$$h\nu\Delta t = \frac{h}{2} \tag{6.5}$$

Thus the time taken for the absorption or emission of a quantum of radiation of energy E is

$$\Delta t = \frac{h}{2E}$$

And so every absorption or emission of a quantum of energy is necessarily a time consuming process. Thus time cannot be separated from any energy transaction. This is an important conclusion that has far reaching implications. One of these is the uncertainty principle, one of the fundamental principles of quantum theory.

6.6   Uncertainty principle

How do we see a material object such as a tree, a bird, or a car? Light falls on the object, bounces off in different directions and enters our eyes to form an image on the retina. So the first step in this process is for light to bounce off the object of our vision.

Let us say we are trying to "see" an electron. Obviously an electron is too small for humans to see. For our eyes to form an image of the object, the object must be extended — it must be large enough. But we can try to measure the essential properties of an electron such as its charge, mass, position, speed, etc. Now, an electron's motion is dictated by two different kinds of force: the gravitational force due to the mass of the electron, and the electromagnetic force due to its charge.[1] The gravitational force is too weak for us to make any observations on a single electron. Electrons set free from conductors in the laboratory fly too fast to be perceptibly affected by gravity. So our observations must be made via the electromagnetic field. And the interaction of the electron with the electromagnetic field takes place through an interaction with the quanta of the electromagnetic field, which are the photons. So when we make a measurement on an electron we need to bounce at least one photon off the electron. Let us see what happens when a photon bounces off an electron. We saw that this process — the Compton effect — involves an absorption and an emission of a photon by the electron. According to Eq. (6.5) the process of absorption or emission of a photon is not instantaneous, but takes an interval of time given by $\Delta t = 1/2\nu$.

If the process of measurement of an electron is done by interacting an electron with a photon, there is bound to be some uncertainty in the results. In absorbing a photon of energy $h\nu$ the energy of the electron increases by $h\nu$ (and in emitting a photon of energy $h\nu$ the energy of the electron decreases by $h\nu$). So let us say we fire a photon of energy $h\nu$ at an electron of energy E. The electron absorbs this photon and so its energy increases to $E + h\nu$. So during the process of absorption the energy of the electron changes by $h\nu$. So we cannot tell what is the exact instantaneous energy of the electron. The process of measurement disturbs the quantity being measured. But we can say that there is an uncertainty in the energy of the electron which is of the order of the energy of the photon it absorbs. So we write:

$$\Delta E \sim h\nu$$

where the wave $\sim$ means "of the order of".

Since this process of measurement takes a certain amount of time which we shall call Δt we can say that within the uncertainty of time Δt there is an uncertainty of energy $\Delta E \sim h\nu$. Now the uncertainty of time is of the order of the time taken to absorb the radiation $\Delta t = 1/2\nu$. So the uncertainty in the energy of the electron and the uncertainty of the time at which the energy was measured bear this relationship:

$$\Delta E \Delta t \sim \frac{h}{2}$$

---

[1] An electron also experiences a third force called the *weak nuclear force* or simply *weak force*. This force is now believed to be an aspect of the electromagnetic force.

But there are different ways of defining uncertainty, some more useful than others. If we define uncertainty using a standard statistical definition the product of the uncertainty of the energy of the measured electron and the duration of measurement (which is the uncertainty in time) has the relationship

$$\Delta E \Delta t \sim \frac{h}{4\pi} \tag{6.6}$$

More accurately, what this equation means is that the *minimum* uncertainties in the two quantities E and t are related as above. The actual uncertainties due to experimental errors, etc. can be much higher. So it is more common to see the uncertainty principle stated as

$$\Delta E \Delta t \gtrsim \frac{h}{4\pi} \tag{6.7}$$

As the photon is absorbed by the electron the former communicates not only its energy but also its momentum to the electron. This means that during this interaction there is an uncertainty in the momentum of the electron. Also, because of the momentum imparted, the electron also gets displaced, which means there is a corresponding uncertainty in its position. Let us label the direction of motion of the incident photon as the x axis. So during the time of interaction Δt, the electron has an uncertainty of position Δx and a corresponding uncertainty of momentum Δp_x. The photon gives an impulse $\mathbf{I}$ ($= \mathbf{F}\Delta t$ where $\mathbf{F}$ is the force applied by the photon on the electron) to the electron which is related to its change of momentum ($\Delta \mathbf{p}$) by the Impulse Momentum Theorem[2]:

$$\mathbf{I} \equiv \mathbf{F}\Delta t = \Delta \mathbf{p} \tag{6.8}$$

So for motion in the x direction

$$F_x \Delta t = \Delta p_x \tag{6.9}$$

where the x subscripts indicate that we are considering motion only along the x direction. Multiplying both sides of Eq. (6.9) by Δx we obtain

$$F_x \Delta x \Delta t = \Delta p_x \Delta x \tag{6.10}$$

Work done on an object = energy communicated to that object (by the work energy theorem, cf. Ch. 2). And work done by a force is defined as the product of the force and the displacement of the object in the direction of the force. The incident photon is traveling in the x direction. So it

---

[2]The Impulse Momentum Theorem follows from Newton's Second Law stated as Force = change of momentum divided by time.

displaces the electron in the x direction and applies a force of F_x in the x direction. So, because of this displacement there is some work done on the electron by the force. Since there is an uncertainty in the displacement, there is a corresponding uncertainty in the work done. The uncertainty in the work done on the electron $= F_x \Delta x$. The work done by the force on the electron $=$ the energy gained by the electron. So, during this transaction, the electron has an uncertainty of energy given by

$$\Delta E = F_x \Delta x$$

Using Eqs. (6.7), (6.9) and (6.10) we obtain

$$\Delta p_x \Delta x \gtrsim \frac{h}{4\pi} \qquad (6.11)$$

The quantity $h/2\pi$ occurs so commonly that it is abbreviated as $\hbar$ (pronounced H bar). So we write the uncertainty principle as

$$\Delta p_x \Delta x \gtrsim \frac{\hbar}{2} \qquad (6.12)$$

Thus, when we try to measure both the x position and the x component of the momentum of an electron using a photon we find that there is a minimum uncertainty in each of these quantities and that the product of these uncertainties cannot be less than of the order of $\hbar$. (A factor of 2 does not affect the order of magnitude.) Conversely, if we picture the interaction between the electron and the photon as an attempt by the electron to measure the position and momentum of the photon we would obtain an identical equation for the photon. By the law of conservation of momentum, whatever be the change in momentum of the electron during the interaction, the photon will undergo an equal and opposite change of momentum. So during the process the change of momentum of the photon will be $\Delta P_x = -\Delta p_x$. If the relative position of the electron with respect to the photon changes by Δx, the relative position of the photon with respect to the electron will change by $-\Delta x$. So we can write for the uncertainty in momentum and position of the photon

$$\Delta P_x \Delta x \gtrsim \frac{\hbar}{2} \qquad (6.13)$$

Heisenberg showed that this uncertainty principle applies to any sort of measurement of an object by any possible method. In fact, one could say that all of quantum theory is about the uncertainty principle. But when we recall that Planck's Constant is an extremely small number, equal to 6.6×10^{-34} in SI units, this uncertainty principle is totally irrelevant for macroscopic objects for which the uncertainties in position and momentum are so great that the inequality (6.12) is easily satisfied by the errors in measurement, without making any appeal to quantum theory.

In the usual mathematical derivation of the uncertainty principle followed by most textbooks the uncertainties are defined in the following manner.

Suppose a number of independent measurements are made of the position of the particle. Let us say we obtain the values x_1, x_2, x_3, etc. The uncertainty Δx is defined as the *standard deviation* of these numbers. So if the average value of x is written as $\langle x \rangle$, then the standard deviation is obtained by taking the square of the deviation of each actual value of x from the mean, taking the average value of this quantity, and then taking the square root. So if there are n measurements made of the position x, then

$$\Delta x = \sqrt{\frac{1}{n}\left[(x_1 - \langle x \rangle)^2 + (x_2 - \langle x \rangle)^2 + (x_3 - \langle x \rangle)^2 + ...(x_n - \langle x \rangle)^2\right]}$$

Exercise 6.5. A charged particle was detected using a screen that produces an image at the spot where the particle struck the screen. The image was examined closely and found to have a depth (in a direction which we shall label as x) of 1.2×10^{-12} m. Taking this to be the uncertainty in the x position of the particle at the moment of its detection, what is the uncertainty in the corresponding momentum p_x?

6.7 Summary

Energy can be absorbed or emitted only in quanta. So an electron interacting with an electromagnetic field cannot exchange an arbitrary amount of energy with the field. The energy exchanged must be $h\nu$ where h is Planck's constant and ν is the frequency of the radiation absorbed or emitted by the electron. This has some important consequences.

Photons are absorbed by electrons and in that process the electrons gain energy. One example of this is the photoelectric effect. When light falls on certain metallic surfaces electrons receive energy from the light photons and for a sufficiently high photon energy — corresponding to a sufficiently high frequency — the electrons gain enough energy to escape the forces that keep them confined to the metal.

A photon can strike a slow moving electron and communicate momentum to the electron. The electron absorbs the photon and almost immediately emits another photon of less energy and therefore lower frequency or longer wavelength. This process of absorption immediately followed by an emission is also called a scattering. This particular form of scattering is called Compton scattering or the Compton effect.

Every interaction between an electron and a field takes a finite amount of time, which is the time taken for the process of absorption or emission of the quantum by the electron. Thus, it can be shown that there is a mathematical relationship between the amount of energy exchanged and the duration of the interaction. This can be expressed as the uncertainty principle

$$\Delta E \Delta t \gtrsim \frac{\hbar}{2}$$

A related uncertainty principle relates the uncertainty in position to the uncertainty in the momentum in the direction in which the position is measured. So if we are measuring the position and momentum in the x direction this uncertainty principle can be written as

$$\Delta p_x \Delta x \gtrsim \frac{\hbar}{2}$$

Chapter 7

Matter Waves

7.1   De Broglie's hypothesis

One of the rules of Einstein's Theory of Relativity is that photons always travel at speed c, whereas electrons and other particles travel at varying speeds but less than c. We will explore some important consequences of this theory in the following chapter. For the present we observe that notwithstanding this difference between photons and other particles, there are important similarities between these two classes of objects. It is these similarities that will be the focus of this chapter. We saw earlier that a photon of frequency ν and wavelength λ has an energy $h\nu$, a momentum $p = h\nu/c = h/\lambda$ and a mass equal to $h\nu/c^2$. So a photon's energy E is related to its mass m by the equation $E = mc^2$. According to Einstein's Special Theory of Relativity the energy E of any object is related to its mass m by the same formula $E = mc^2$. In the previous chapter we saw that the uncertainty principle applies equally to an electron and a photon. So it appears that there are some important similarities between photons and material particles. Photons are quanta of the electromagnetic field. Could it be that electrons are also quanta of some field? If they are the quanta of some "electron field" then we would expect electrons also to be propagated as waves. We have stated above that the momentum of a photon is related to its wavelength by the formula $p = h/\lambda$. So if an electron has a momentum of magnitude p, we would expect it also to have a "wavelength" λ that is related to p by the formula

$$p = \frac{h}{\lambda}$$

This is the hypothesis put forward by Louis de Broglie. This is a revolutionary hypothesis with far reaching consequences. De Broglie speculated that every material particle — be it an electron, a proton or a neutron —

has a corresponding wavelength. A particle having momentum mv would have wavelength

$$\lambda = \frac{h}{mv}$$

This wavelength is called the *de Broglie wavelength* of a particle. So de Broglie suggested that every material object is both a particle and a wave. This has become the foundational concept of quantum theory. But at the time when de Broglie put forth this revolutionary thesis, neither he nor anyone else knew what it meant to say was both a particle and a wave. But as quantum theory developed through the ensuing decades, a picture began to emerge that provided a sort of definition of the wave-particle duality of quantum theory. Wave-particle duality in quantum theory can be described in two statements:

1. Photons, electrons and other material entities *propagate as waves* — they undergo interference according to the properties of waves.

2. Photons, electrons and other material entities are *detected as particles*. When they are "caught," they are found to have mass, momentum, and other properties traditionally associated with particles. We shall next explore the phenomenon of interference and show how this illustrates the "wave" aspect of both photons and electrons and how the detection of photons and electrons on a screen illustrates their "particle" aspect.

7.2 Interference

7.2.1 *Combination of waves*

When a wave travels through a medium, the particles of the medium vibrate, and the shape of the vibration is propagated along the medium. So the energy of vibration is carried along by the medium, though the particles of the medium remain in the same locality as they vibrate about a fixed position.

Consider the following wave:

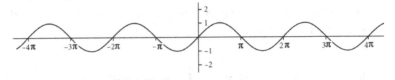

The graph shows the shape of the wave at a particular instant of time. The vertical axis y is along the direction of displacement and the horizontal axis x is along the direction of propagation. For this wave the wavelength — the distance between two successive crests (or two successive troughs) — is $\lambda = 2\pi$ meters. The maximum displacement or amplitude $A = 1$ meter.

Next we examine the following wave:

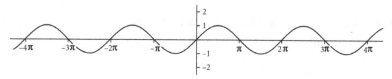

The two waves are identical. Let us see what happens when both the waves are superposed — i.e. placed on top of each other or placed together. The two waves are exactly in phase, i.e. the crests of one wave coincide with the crests of the other and the troughs of one wave coincide with the troughs of the other. The result will be that the particles of the medium will undergo a displacement that is the sum of the displacements of the individual waves. So if the displacement due to the first wave is y_1 and that due to the second wave is y_2 the resultant displacement will be $Y = y_1 + y_2$. The shape of the wave is as follows:

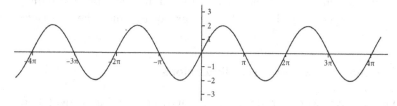

We see that the wavelength remains the same. But the amplitude is the sum of the individual amplitudes, equal to $1 + 1 = 2$ meters. This is what happens when the two waves are exactly in phase. When the two waves are exactly in phase, we say that there is **constructive interference** between the component waves.

Next, we consider two waves that are exactly out of phase, so that the crest of one wave coincides with the trough of the other wave:

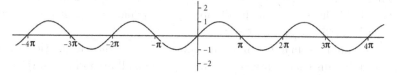

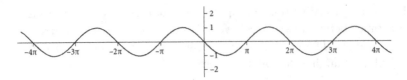

When these waves are superposed on each other, the result is that there is no displacement anywhere, no energy and no wave:

When the two waves are exactly out of phase, we say that there is **destructive interference** between the waves.

In these two examples we considered waves with equal amplitudes that are either exactly in phase or exactly out of phase. But it is possible that the two component waves may not be exactly out of phase. In such a case the resultant amplitude of the wave will be diminished, but not reduced to zero. The amplitude will depend on the extent to which the two waves are out of phase.

7.3   Waves in two or three dimensions

So far we have been considering waves moving along a straight line, i.e. one-dimensional waves. But real waves move through real media, which are three-dimensional.

If we drop a stone into a pool of water, waves move outwards from the spot of the disturbance. Since the speed is the same in all directions, the waves at the surface of the water take on a circular shape with the disturbance at the center. We say that these waves have a *circular wave front*.

The light from a candle travels outwards three dimensionally, and since the speed of light is the same in all directions, the light has a *spherical wave front* with the flame at the center.

Light coming from a distant object like a star does have a spherical wave front, but because the center is so far away, a small area of the surface of the sphere is like a plane, and so this light has a *plane wave front*. If now we pass the light from a star through a narrow opening in the shape of a slit, the light emerging on the other side of the slit would have a *cylindrical wave front*.

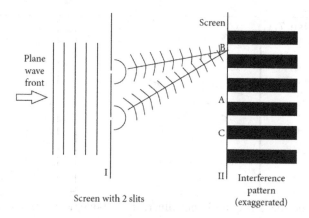

Screen with 2 slits

Interference pattern (exaggerated)

In the figure above the source is kept far to the left of screen I. So the light reaching screen I has a plane wave front. This wave is blocked by the screen except for a small amount that is allowed to pass through the two slits (cut perpendicular to the plane of the diagram). These two slits behave like sources producing waves with a cylindrical wave front. The cross-section shown in the diagram has circular wave fronts. These two waves proceeding from the two slits will undergo *interference*. Consider a point A which receives light from both the slits. If the distance of A to the upper slit is equal to the distance of A to the lower slit, then both the light waves will reach A in phase. Thus there will be a large amplitude at A, and so the intensity of light will be great at A. So there will be a bright spot at A. Suppose now there is a point B such that the difference in the distances of the two slits from B is equal to a whole number of wavelengths. Then the two beams will arrive at the point B in phase. There will therefore be a bright spot also at B. Next consider a point C such that the difference in the distances between C and the two slits is an odd multiple of half a wavelength, so that a crest of one wave and a trough of the other wave reach C simultaneously. The two waves will be exactly out of phase when they arrive at C. So there will be destructive interference and hence a dark spot at C. But since the openings on screen I are not tiny holes but narrow slits, the wave fronts between screens I and II are cylindrical and so the pattern we find on screen II would not be dots of light but alternate bright and dark bands perpendicular to the plane of the figure. This pattern is shown on the extreme right in the diagram. Such a pattern is called an interference pattern. An interference pattern demonstrates that light travels in the form of waves.

Exercise 7.1.

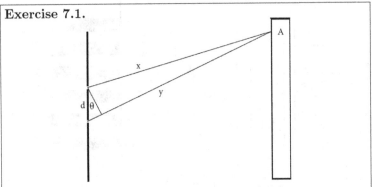

In the double slit experiment illustrated above, the slits are separated by a small distance d. The lines labeled x and y represent the paths taken by light from the two slits reaching the point A on the screen. The light waves are in phase when they leave the slits. But because of the difference in distance traveled, they may not reach the screen in phase, unless their path difference equals a whole number of wavelengths. Suppose the angle θ is sufficiently small, and the separation between the screen with the slits and the image screen is sufficiently large, so that the difference between x and y can be taken to be the side of the little triangle opposite the angle θ.

Show that the condition for a bright spot at A is given by

$$\sin \theta = \frac{n\lambda}{d}$$

where n is a whole number.

The double slit experiment was originally carried out by Thomas Young who proved that light is a form of wave motion, and not a stream of particles, as Newton had suggested. Nevertheless, the quantum theory has now shown that Newton was not totally wrong after all. In the following section we shall see how the double slit experiment has been refined to illustrate the particle or quantum nature of light.

7.4 Quantum theory of light

Suppose we were to reduce the brightness of the source until it is so feeble that it cannot be seen with the human eye. If we now replace the screen II by a photosensitive plate of very high resolution, and mount a powerful microscope on the other side of the plate, we would observe an interesting thing.

Initially the photosensitive plate would be dark. Then suddenly a single bright spot would appear in a region where a bright band used to be, and later another bright spot where another bright band used to be when the source had full power. A little later a bright spot would appear in line with the first bright spot — within the same band as one of the original bright bands observed earlier — and another bright spot where another bright band used to be, and so on. After a long time, there will be millions of spots which will eventually merge to form the interference pattern that we saw when a bright source was used.

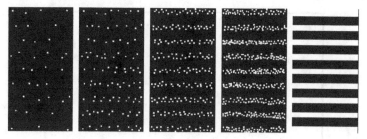

Each bright spot represents a unit of light energy — a quantum, or a photon. This experiment shows that light energy is not emitted or absorbed continuously, but only in drops or units. What we have done in this experiment is to lower the intensity of the source so much that only about one quantum of light energy — or one photon — is emitted per second. This experiment proves that light energy is quantized, that it exists only in discrete multiples of a basic unit of energy.

This then raises a question. If only one photon were traveling from the source to the screen II, through which slit did it travel across screen I — the upper or the lower slit? The surprising answer is that the photon went through both slits, or, more precisely, electromagnetic energy traveled through both slits, and this energy was detected somewhere along screen II as a single photon.

The electromagnetic wave went through both slits, and became two waves on the other side of screen I. The interference between these two waves determined where the photon would be found on a detector such as screen II. But we saw that there is a certain randomness to the process. Each individual photon did not seem to follow any set procedure for falling on the screen, except that it landed only where the bright bands appeared when the source had full intensity.

Suppose we were to keep track of all the light quanta by following the appearances of all the spots of light at different points on the screen. So we could draw numbers indicating the order in which the spots of light appeared at different places on the screen, 1 for the first spot that appears, 2 for the second spot, etc. A typical pattern with numbered spots may have 1 at the mid point or center, 2 at a point some distance below 1, 3 above 1, 4 close to 1, etc. After a long time, the spots will be clustered so closely that they are no longer discernible as individual spots but merge into the interference pattern we saw earlier.

23 6		10	17		19	
8	11	15	22 3		14	21
18	4	1	25 7		12	20
2	9	16	24 5		13	

Then we switch off the source, remove the photographic plate, and replace it with a fresh plate. Now we switch on the source as before, and monitor the appearances of the spots of light. This time the order in which the spots arrive will not be the same as the first time. But after a long time, after millions of spots have appeared, the same interference pattern will emerge as before. This shows that while the electromagnetic wave determines all the possible locations where the light energy could be found, it does not tell any one quantum of light energy where to go. There is a randomness in the way that the individual light quanta "decided" where to land on the screen.

```
            22      14              2    18       24
       11
   17       23              9            12    6
                   15
   20     1   5      10         19     3        13

        7     21        8       16     4     25
```

The regular pattern did not appear with one or two photons, but with billions of them. So the interference pattern is a statistical effect. The apparently well-defined image is formed by billions of spots appearing at random but falling in preferred locations. The probability for the appearance of a spot is determined by the geometry of the apparatus, but an individual photon is free to choose any location that is permitted by the probability. Over the long run the number of photons arriving at any one location is proportional to the probability for a single photon to arrive at that location.

Now, the refraction of light through a lens is ultimately due to wave interference, even though elementary textbooks treat this subject as "geometric optics." What this means is that the image produced in our retina is due to the impression created by billions of photons arriving at random following the laws of probability as they pass through the lens in our eye. Our perception of reality is therefore a statistical effect. Maxwell had correctly stated that "the true logic of this world is in the calculus of probabilities."

7.5   Electron waves

De Broglie suggested that every material particle has a wavelength related to the momentum of the particle by the formula $\lambda = h/p$. Planck's constant is a very small number, equal to 6.6×10^{-34} in the usual SI system of units. The mass of an electron is 9.1×10^{-31} kg. So for an electron traveling at the speed of 1 m/s, the wavelength would be 0.000725 m = 0.725 mm, which is close to a millimeter. Thus a slow moving electron will propagate as a wave with a wavelength of about a millimeter. So if we could produce such slow moving electrons, we could pass them through a double slit and observe the interference pattern on a screen. But such slow moving electrons cannot be found moving through a vacuum. As we explained earlier in our discussion on the photoelectric effect, electrons cannot escape from a metal — and thus be noticed — unless they are given a great deal of energy. The typical speed of an electron that is accelerated through a vacuum is about a million meters per second. Since the momentum is very high, the wavelength is very short. The only way to observe interference is to have slit separations of the order of the separation between atoms in a metal.

Such an experiment was indeed carried out by Davisson and Germer who showed that the electron waves underwent interference when passing through a nickel plate. This sort of interference — between a very large number of waves — is called *diffraction*. An independent experiment was successfully performed by G. P. Thomson. Thus de Broglie's hypothesis was established as a law of physics.

7.6   Composite particles

We have so far been considering two kinds of particles: photons, which are the quanta of light and other forms of electromagnetic radiation, and electrons, which are negatively charged particles. Both these particles are fundamental particles, that is, they cannot be separated into more elementary particles. But not all particles are fundamental. A hydrogen atom consists of a proton and an electron, attracted by each other's opposite charges. We might call a hydrogen atom a composite particle, made up of two or more fundamental particles. We now pose the question: does a composite particle such as an atom also have a de Broglie wavelength? The answer is yes.

Let us consider this hydrogen atom to be moving with the speed v.

According to de Broglie's formula this hydrogen atom would have a wavelength

$$\lambda_H = \frac{h}{(m_e + m_p)v}$$

(The mass of particles moving with speeds approaching c would be greater than that of the same particle moving with a slower speed. Thus the expression inside the brackets in the above equation would have to be modified for very fast moving particles. We shall discuss the mechanics of very fast motion in the next chapter.)

Microscopic particles entering into a partnership with one another is called *entanglement*. Such a partnership will last even when the individual particles are separated by large distances.

Entanglement can occur if the two particles are of the same type — called *identical particles* — or of different types, such as the example given above. Entanglement can also occur between a very large number of particles, such as the carbon atoms in a diamond, or the iron atoms in a nail, etc. So each of these collections of particles behaves as a single particle in having a distinct wavelength.

Let us find the de Broglie wavelength of a baseball traveling at the relatively slow speed of 10 m/s. The mass of a baseball is about 0.14 kg. So the wavelength of a baseball moving at 10 m/s would be 4.7×10^{-34} m. This is an extremely short wavelength, much smaller than the size of a hydrogen atom. When the wavelength is very short, it is hard to see interference effects. So for the double slit experiment, the separation between the bright and dark bands on the screen would be extremely short for very short wavelengths. If the wavelength is only about 4.7×10^{-34} m, it would be impossible to tell the dark bands from the bright bands since they would be too close together. So it is impossible to show experimentally that a baseball propagates as a wave.

Exercise 7.2. Find the de Broglie wavelength of (a) a proton traveling at a speed of 8.00×10^3 m/s given the mass of a proton $= 1.67 \times 10^{-27}$ kg and (b) a Helium 4 nucleus of mass 6.64×10^{-27} kg traveling at the same speed. Take $h = 6.63 \times 10^{-34}$ in SI units.

7.7 The hydrogen atom

The importance of de Broglie's formula for the wavelength of a particle became evident when he offered a brilliant explanation for one of the puzzles of that time. This puzzle concerned the structure of atoms. By now it was known — thanks largely to J. J. Thomson and E. Rutherford — that atoms consist of a positive nucleus that contains the bulk of the mass of the atom, and negative electrons that are apparently orbiting the nucleus like planets around the sun. The simplest atom is the hydrogen atom, and so, physicists were trying to understand how this simplest atom functioned before they could move on to more complex atoms. At this point it was clear that infrared, visible, and ultraviolet radiations were emitted by atoms, and also absorbed by atoms, and that it was the electrons in the atom that were responsible for these transactions. We saw earlier in Ch. 4 that an electric charge that is oscillating or vibrating emits electromagnetic waves. A charge moving in a circle would also emit such waves according to classical electromagnetism. However, that does not happen in the case of atomic electrons. Atoms are stable. Moreover, the radiation emitted or absorbed by atoms appears only with certain frequencies, and these frequencies seem to have some kind of mathematical relationship with each other, which cannot be explained by classical physics. The relationship that was discovered from experiment was the following:

$$\text{Frequency of the radiation} = \text{Constant} \times \left(\frac{1}{n_1^2} - \frac{1}{n_2^2} \right)$$

where n_1 and n_2 are natural numbers 1, 2, 3, ...etc. with $n_2 > n_1$.

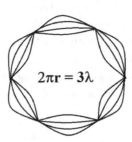

Fig. 7.1. Atomic orbitals as standing de Broglie waves. Here three wavelengths are contained in one orbit, corresponding to $n = 3$. There are six nodes (zero displacement) and six antinodes (maximum displacement).

Niels Bohr had proposed that there are certain stable orbits in which the electrons do not radiate electromagnetic waves, and that radiation is emitted only when an electron drops from a higher orbit (corresponding to the number n_2) to a lower orbit (corresponding to the number n_1).

Likewise, an electron in a lower orbit could absorb radiation and jump to a higher orbit. Bohr suggested that the radius of a stable electron orbit should be proportional to n^2, where n is a natural number, 1, 2, 3, ... This hypothesis has an important consequence. Every object moving in a circular path has angular momentum equal to the radius of the path multiplied by the momentum (mass multiplied by velocity) of the object. A detailed calculation shows that Bohr's hypothesis implies that in a stable orbit the angular momentum of the electron due to its orbital motion should be a whole multiple of $\hbar$. So it can be fairly stated that Bohr's hypothesis regarding the radius of atomic orbits is equivalent to saying that the angular momentum of an orbit is a whole multiple of Planck's constant. On the basis of his hypothesis Bohr was able to calculate the energies of the permitted orbits. He then subtracted the energy of a lower orbit (n_1) from the energy of a higher orbit (n_2) and showed that this difference $E_2 - E_1 = h\nu$ where ν is the frequency of an observed radiation which was experimentally shown to belong to the pattern

$$\text{Constant} \times \left(\frac{1}{n_1^2} - \frac{1}{n_2^2} \right).$$

Bohr was clearly on the right track, since his hypothesis helped him arrive at an experimentally observed result. But this prompted the question: why should these orbits with angular momentum multiples be the only permitted ones? And to this de Broglie provided the answer. We saw that according to de Broglie every particle of momentum $p = mv$ has a wavelength associated with it equal to $\lambda = h/p$. He now suggested that the permitted electron orbits were the circular standing waves of the electron. His explanation was that when an electron orbits the nucleus the circumference of the orbit should be equal to an integer multiple of the wavelength of the electron. See Fig. 7.1. In other words, the paths of the orbiting electrons should be thought of as stationary waves, similar to the standing waves in a vibrating string. The difference is that in a vibrating string the waves are linear. Here, the waves move along a closed circle.

Using this principle, if the radius of the nth orbit is r, we would have

$$2\pi r = n\lambda = \frac{nh}{mv} \tag{7.1}$$

So not all radii are permitted, but only those that have a specific relationship to the speed v of the electron:

$$vr = \frac{nh}{2\pi m} \tag{7.2}$$

This equation may be rewritten as

$$mvr = \frac{nh}{2\pi} = n\hbar \qquad (7.3)$$

The quantity on the far left is the angular momentum of the orbiting electron. This angular momentum is seen to be a whole multiple of $\hbar$. This is in full agreement with Bohr's condition for a permitted orbit, but now we see that this condition appears as a logical consequence of de Broglie's fundamental law. Bohr's hypothesis was *ad hoc*, put forward to explain a specific phenomenon. De Broglie's law is universal. So the *ad hoc* was now explained as a particular consequence of the universal.

Now, an electron in orbit round the nucleus is kept in a circular orbit by the centripetal force which is the electrical force of attraction between the electron and the proton (which constitutes the nucleus of a Hydrogen atom). According to the laws of electromagnetism this force is given by

$$\frac{e^2}{4\pi\epsilon_0 r^2}$$

where ϵ_0 is the permittivity of free space.

This centripetal force generates a centripetal acceleration. And the centripetal acceleration of a particle moving with speed v along a circle of radius r is given by (Ch. 2):

$$\frac{v^2}{r}$$

By Newton's second law (mass × acceleration = force),

$$\frac{mv^2}{r} = \frac{e^2}{4\pi\epsilon_0 r^2} \qquad (7.4)$$

By combining Eqs. (7.2) and (7.4) we obtain an expression for the radius

$$r = \frac{n^2 h^2 \epsilon_0}{\pi m e^2} \qquad (7.5)$$

Exercise 7.3. Find the radius of the lowest orbit for a hydrogen atom using Eq. (7.5). This radius is called the *Bohr radius*. Check your answer with a reliable online source.

The total energy of an atom = kinetic + potential. The kinetic energy of the atom is entirely due to the orbital motion of the electron, since the nucleus (which is the proton for the hydrogen atom) can be taken to be stationary in comparison with the fast motion of the electron round the nucleus. The potential energy of the atom is the potential energy due to the electrical force between the electron and the nucleus. So the total energy

$$E = \frac{1}{2}mv^2 - \frac{e^2}{4\pi\epsilon_0 r} \tag{7.6}$$

The potential energy is negative because the force between the electron and nucleus is attractive.

Substituting expressions for v and r obtained earlier, and using some simple algebra, we get

$$E = -\frac{e^4 m}{8\epsilon_0^2 h^2 n^2} \tag{7.7}$$

We find that the total energy is negative. This means that the electron cannot escape from the nucleus unless it receives some energy from outside. If this positive energy given to the electron is greater in magnitude than the negative energy of the atom that would enable the electron to break free from the nucleus. But if it receives energy somewhat less than this amount then it may be possible for the electron to jump to a higher orbit with a higher value of n. Normally the electron occupies the lowest possible energy state, for which $n = 1$. If it receives some energy, it could jump to an orbit where the value of $n = 2$, 3, 4, etc.

The lowest possible energy state of an atom is called its *ground state*. For a Hydrogen atom the ground state is the energy state with $n = 1$. States with higher values of n are called *excited states*. When an atom is in an excited state it would normally return to the ground state by emitting a light particle or photon. The energy of this photon would be equal to the difference in energy between the excited state and the ground state of the atom. This energy is given by

$$E_2 - E_1 = \frac{e^4 m}{8\epsilon_0^2 h^2} \left(\frac{1}{n_1^2} - \frac{1}{n_2^2} \right) \tag{7.8}$$

If λ is the wavelength and ν the frequency of the emitted photon, we can write

$$E_2 - E_1 = h\nu = \frac{hc}{\lambda} = \frac{e^4 m}{8\epsilon_0^2 h^2} \left(\frac{1}{n_1^2} - \frac{1}{n_2^2} \right) \tag{7.9}$$

The experimental physicists used a quantity called *wave number*, which is the reciprocal of the wavelength:

$$\frac{1}{\lambda} = \frac{e^4 m}{8\epsilon_0^2 h^3 c} \left(\frac{1}{n_1^2} - \frac{1}{n_2^2} \right) = R \left(\frac{1}{n_1^2} - \frac{1}{n_2^2} \right) \tag{7.10}$$

where R is called the Rydberg constant, which was originally obtained from experiment. The value of R may be calculated theoretically from the formula for R in the above equation. It works out to be 10 973 731.6 m^{-1} and agrees perfectly with the experimental value.

> **Exercise 7.4.** Calculate the theoretical value of the Rydberg constant. If you use good approximations such as taking $e = 1.6 \times 10^{-19}$ you will get a value very close — but not exactly equal — to the number given above.

7.8 Summary

We saw in Ch. 5 that the electromagnetic radiation in a Black Body Cavity existed as standing waves where the frequencies of the radiation had a definite relationship to the dimensions of the cavity. Applying the principle of equipartition of energy to all the possible modes led to the ultraviolet catastrophe which required a drastic quantum solution. This was the beginning of the quantum theory of radiation that led to our understanding of light and other forms of electromagnetic radiation as consisting of discrete quanta or photons.

In the case of electrons we began with the historical assumption that they were particles with definite mass, momentum and spatial location. Louis de Broglie showed that electrons and other "material" particles also have wavelength that is related to the momentum of the particle by the same formula as for a photon, viz. $p = h/\lambda$.

De Broglie was able to show that the motion of the electrons in an atom could be explained by assuming that the orbits of the electrons are standing waves with the circumference of the orbit equal to a whole multiple of the wavelength of the electron. Using this principle he calculated the total energy of an atom. The wave nature of electrons was verified directly by observing the diffraction of a beam of electrons through a piece of metal.

Thus the quantum theory has served to unify two entities that were hitherto considered quite distinct — matter and waves. Electrons were thought of as matter, and light was thought of as a non-material wave. Now we see that such a distinction between electrons and photons is untenable.

The quantum theory holds for all kinds of particles besides electrons and photons. De Broglie's Law and the Uncertainty Principle apply to protons, neutrons, muons, mesons, neutrinos and every possible fundamental particle. They also apply to groups of particles such as atoms and molecules.

Chapter 8

The Special Theory of Relativity

8.1 Speed of light

We saw earlier that a vibrating charge sends out electrical influences as waves which in turn create magnetic influences which also travel along with the electric influences as waves (Ch. 4). This combination of electric and magnetic waves is called an electromagnetic wave, and all electromagnetic waves travel at the same speed. Light is a form of electromagnetic wave. We have also seen that these electromagnetic waves carry energy in the form of packets called photons.

In this chapter we are concerned with speeds. An important question concerns the speed of light as measured by an observer moving relative to the source. Will the speed of this light as seen by the observer be greater than 3×10^8 m/s if the observer is approaching the source, and less than this quantity if the observer is receding from the source? This is actually a far more serious question than might appear, and we shall take it up in the following sections.

8.2 Relative speed in classical mechanics

Suppose two cars A and B are approaching each other. A is traveling at 20 m/s and B at 15 m/s. What is the relative speed of A with respect to B? The classical answer is $20 + 15 = 35$ m/s. The basic rule is to add the two speeds.

Let us next consider two cars A and B moving in the same direction. A travels at 12 m/s and B at 20 m/s. What is the relative speed of B with respect to A? The answer is $20 - 12 = 8$ m/s.

In general, when two objects approach each other along the same line with speeds u and v, their relative speed is $u + v$. And if car A moves with speed u and car B with speed v in the same direction as A, then the relative speed of B with respect to A would be $v - u$, and this speed would be positive if B is faster than A and negative otherwise. Here positive means B is moving further away from A, and negative means A is catching up on B and will eventually pass B.

Suppose now a car is moving with speed u towards a source of sound. Let the speed of sound in air be w. How fast will the sound be traveling relative to the car? The answer is $u + w$.

Next, we consider a car moving with speed u receding from a source of sound. How fast will the sound be traveling relative to the car? The answer is $w - v$.

In both the last two cases we have assumed that the car is not supersonic, that it is traveling at ordinary speeds — much slower than sound.

So the measured speed of sound will be different from the actual speed of sound in air when the observer is in motion either towards or away from the source of sound. (But if the observer is stationary, the speed of sound will remain the same regardless of the speed of the source, since the sound waves are carried by the medium of air, and the speed of sound waves depends only on the medium, not the motion of the source. To be sure, if there is a wind blowing from the source to the observer, then the observer will detect a greater sound speed.)

8.3   Motion relative to source of light

What happens if a moving observer attempts to measure the speed of light? Would the speed of light appear greater if the observer is moving towards the source of light? And would the speed of light appear lower if the observer is moving away from the source of light?

An experiment was carried out in 1887 by Michelson and Morley which basically attempted to answer the above two questions. The upshot of their experiment is that there is no difference in the speed of light as measured by an observer moving in the same direction as the light, and by an observer moving in the opposite direction as the light.

Different explanations were given for the negative results obtained by Michelson and Morley. Most of the explanations were partially right, but none of them gave the full picture, until Einstein provided the complete and correct explanation 18 years later.

8.4 Principles of special relativity

Einstein published his Special Theory of Relativity in 1905. The word Special indicates that this theory deals only with objects that are in motion in a straight line with constant speed. Also, the Special Theory does not deal with effects of gravitation. The more general form of Relativity which includes acceleration and gravitation is called the General Theory of Relativity, which Einstein published in 1915.

The two principles of Special Relativity can be summed up as follows:

Principle 1. *The laws of physics are the same when observed within any laboratory that is not accelerating.*

Principle 2. *The speed of light (and other electromagnetic waves) is the same no matter what be the speed of the source, or the speed of the observer.*

Note: A common term for a laboratory from which observations are made on objects both within and without the laboratory is *frame of reference*. Of course, a frame of reference need not be an actual laboratory housed in a building.

Principle 1 implies that all frames of reference are equivalent if they are not accelerating. They can be moving at any speed relative to each other. An experiment would yield the same result no matter which frame it is performed in. This also implies that no experiment can be done entirely inside a frame to determine how fast the frame is moving relative to some outside object. In order to get that information it would be necessary to make observations of events occurring outside the frame.

This principle also implies that there is no such thing as Absolute Rest. We know that the earth is revolving round the sun, and that the sun itself is moving around the center of the Milky Way. Can we leave our galaxy and find some spot where we can be absolutely stationary? That would not help us, because there is no such thing as being *absolutely* stationary. We can only be stationary relative to some other object.

Principle 1 is not really new. Galileo and Newton would have agreed that this principle was true to the behavior of Nature and that it was just a new way of stating the law of inertia.

Principle 2 has no place in classical kinematics. Neither Galileo nor Newton could have dreamed of such a principle.

Using these two principles Einstein made some mathematical calculations and obtained some truly extraordinary results.

8.5   Relative speed according to Einstein

One important consequence of the Special Theory is that the equation for relative speeds is modified. The correct equation for the relative speed w of two cars approaching each other with speeds u and v relative to the ground is no longer $w = u + v$ but

$$w = \frac{u + v}{1 + \frac{uv}{c^2}} \tag{8.1}$$

where c is the speed of light (3×10^8 m/s).

This formula is exactly true for all speeds and for all objects — cars, space ships, planets, stars, elementary particles, atoms, molecules, etc.

For ordinary speeds the relative speed becomes practically the same as $u + v$. For example, if we have two cars approaching each other with speeds of 30 m/s, their relative speed will be

$$\frac{30 + 30}{1 + \frac{30 \times 30}{3 \times 10^8 \times 3 \times 10^8}} = \frac{60}{1 + 10^{-14}} = \frac{60}{1.00000000000001} = 59.9999999999994 \, \text{m/s}$$

This is almost exactly equal to 60. With our present speedometers it is indeed impossible to tell the difference between this number and 60. So at ordinary speeds, the formula for relative velocity can be approximated as $u + v$.

But when we consider speeds approaching the speed of light, the difference between the classical and the relativistic formulas becomes significant.

Let $u = 0.99c$ and let $v = 0.99c$ where c is the speed of light.

The relative speed of two spaceships approaching each other with speeds of $0.99c$ with respect to the earth becomes

$$\frac{0.99c + 0.99c}{1 + \frac{0.99c \times 0.99c}{c^2}} = 0.999949497c$$

The relative speed is very close to c, but is a little less than c. By contrast, the classical formula ($u + v$) would yield a relative speed equal to $1.98c$, which is nearly twice the speed of light. The formulas of the Theory of Relativity make it impossible for the relative speed between two objects to exceed the speed of light.

Suppose now we consider an observer moving at a speed u towards an approaching beam of light. The light approaches at speed c. The speed of this light beam relative to the observer is given by the formula

$$\frac{u + c}{1 + uc/c^2} = c$$

If the observer were receding from the source of light at speed v, he would see the light approaching him with the speed

$$\frac{c - v}{1 - cv/c^2} = c$$

So the speed of light will be the same regardless of the speed of the observer.

8.6 Impossible to attain the speed of light

We can also show that it is impossible for any object to be speeded up to reach the speed of light.

In order to accelerate an object, we have to apply force to it. And force is an interaction between two objects. Object A exerts a force on object B. And in that process object B exerts an opposite force on object A. Jet planes accelerate by pushing the jet of air rapidly in the backward direction. Rockets accelerate by expelling the exhaust with a high force and thereby being pushed forward by the exhaust. Newton's third law is valid in Relativity as well.

Consider the following scenario. A massive space ship takes off vertically and flies at a speed of 0.9c away from the earth. Then this space ship fires a smaller space craft vertically upwards at a speed of 0.9c relative to the big space ship. All the motion is upwards, and so directed away from the earth. Let us neglect the recoil of the big space ship, assuming that the mass of the big ship is much greater than the mass of the space craft. So relative to the large space ship the small space craft is seen to be traveling at 0.9c. The large space ship itself is moving at 0.9c relative to the earth. So the speed of the small space craft relative to the earth is NOT 1.8c as might be expected from classical physics but is actually

$$\frac{0.9c + 0.9c}{1 + \frac{0.9c \times 0.9c}{c^2}} = 0.994475c$$

And this speed is very close to, but definitely less than, the speed of light c.

Now, if we were to consider the acceleration of a rocket as being made up of a very large number of such vertical pushes we would readily see that the final speed of the rocket relative to the earth will always be less than c, even though this speed can come very close to c.

No matter how we try to accelerate an object, we can never make it travel faster than light, or even as fast as light.

The speed of light is the ultimate speed limit of the universe.

Exercise 8.1. Two spaceships are traveling in opposite directions at speeds of 0.99c relative to the earth. Find the speed of the spaceships relative to each other. (This must be less than c.)

8.7 Length is relative

Fig. 8.1.   An object becomes flattened as it moves relative to an observer. The extent of the shortening increases as the relative speed approaches c. Here the sphere is moving in an upward direction. (The flattening is not drawn to scale.)

Suppose we build a space ship of length L_0 from front to back. As this ship is flying at a speed v relative to an observer on earth, this observer will see the length of this ship as somewhat shorter. The exact formula for the length of a moving object is given by

$$L = L_0\sqrt{1 - \frac{v^2}{c^2}} \tag{8.2}$$

Here L_0 is the length of the object when it is stationary, and L its length as measured by an observer relative to whom the object is moving with speed v.

The above formula is true for all speeds. But unless the speed is close to the speed of light the length contraction is negligible.

So consider a train of length 100 m. If this train is traveling at 30 m/s, what will be its length as seen by a stationary (with respect to the ground) observer?

The answer is

$$100\sqrt{1 - \frac{900}{9 \times 10^{16}}} = 99.9999999999995 \text{ m}$$

This is very slightly shorter than 100 m, and the difference is so small as to be undetectable. So for ordinary speeds, length contraction is not an issue. But when speeds begin to approach c, then length contraction becomes significant.

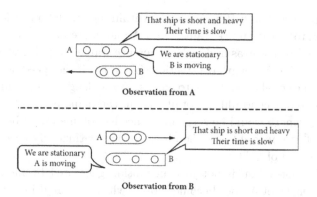

Fig. 8.2.   A and B are spaceships moving in opposite directions relative to each other.

Consider a space ship A of length 100 m traveling at a speed of 0.9c relative to an identical space ship B moving in the opposite direction.

According to A, the length of B will appear as

$$100\sqrt{1 - \frac{0.9c \times 0.9c}{c^2}} = 43.6 \text{ m}$$

Likewise, B will see that A has shortened to 43.6 m.

This is the novelty of Einstein's Theory of Relativity. Ordinarily, when we say that height is relative, we mean that man A is taller than B, and that man B is shorter than A. Also, a person may think she is tall, but a taller person will consider her to be short.

But when we say that length is relative in Einstein's theory, we mean that A is shorter than B when measured by B, and that B is shorter than A, when measured by A.

If we picture A and B as operating within moving laboratories, we call each laboratory a reference frame. So if A and B are in relative motion, A is contracted when measured in B's reference frame, and vice versa.

The equation for length contraction shows that if the speed of the object could become equal to c relative to an observer its length would shrink down to zero. The object would be totally flat in the plane perpendicular to its motion. Its length would be zero, and hence its volume would be zero. But of course it is impossible for the object to be continuously accelerated to reach the speed of light.

Length contraction is not just an illusion generated by motion. It is real. The length of A does become shorter when measured in B's reference frame. And the length of B does become shorter when measured in A's reference frame. And it does not matter what physical apparatus we use to perform that measurement.

8.8   Time ordering of events is relative

Let us imagine two identical trains A and B which have a length of 1000 m when they are stationary.

Now suppose they are traveling in opposite directions at a relative speed of 0.9c.

As they pass each other in parallel tracks, the passengers in A will see B contracted to 436 m, and the passengers in B will see A contracted to 436 m.

Now, suppose the engineer of each train splashes paint on the caboose of the other train as he passes it. Which train will be splashed first?

Event I : Engineer of A splashes caboose of B.

Event II: Engineer of B splashes caboose of A.

Now, train A will see a shortened train B. So the engine of A will pass the caboose of B before the engine of B passes the caboose of A (from A's perspective).

So from A's perspective Event I will take place first. Some time later, the engine of B will have reached the caboose of A and so Event II will take place with the engineer of B splashing the caboose of A.

So as seen by A: Event I takes place first. Event II takes place later.

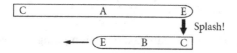

Fig. 8.3.   Event I as seen by A. Event I happens before Event II in this point of view.

Now, let us observe the whole thing from B's point of view. According to B, train A will be shorter than B. So the caboose of A will pass the engine of B first. So the engineer of B will splash the caboose of A first. So Event II occurs first when observed by B. A little later, train A has moved further and the engine of A has passed the caboose of B. So Event I now occurs, with the engineer of A splashing the caboose of B.

So as seen by B: Event II takes place first. Event I takes place later.

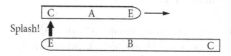

Fig. 8.4.   Event II as seen by B. Event II happens before Event I in this point of view.

Thus we see that the time ordering of two events is not the same when seen by two different observers who are moving relative to each other. Two events observed in different reference frames may not necessarily have the same order in time.

8.9   Duration of time is relative — time dilation

We have seen that length — which is an interval in space — is relative. We also saw that the time ordering of events is relative. Einstein's calculations showed that duration of time — or time interval — is also relative.

Let us consider the two trains discussed in the previous section.

Let T_0 be the time taken for some event to occur on train A as observed by passengers traveling on A. When the same event is observed by passengers on B they will find that this event took a longer time. So if they watch the event happening on A and time it by their clock they would find the duration T of that event as

$$T = \frac{T_0}{\sqrt{1 - \frac{v^2}{c^2}}} \tag{8.3}$$

So let us suppose there is a card game match going on inside train A. Let us say that according to the passengers on A, this match took 100 minutes to complete. But according to the passengers on B who are watching the events on A, the duration of the same match was 229 minutes.

This means that when passengers in B are watching A, they will find that everything in A is moving slowly. So B will observe that time is moving slowly on A.

Similarly, when passengers in A are watching B, they will find events in B occurring slowly, and so they conclude that time is running slowly on B. This phenomenon is called *time dilation* or *time dilatation*.

So time is relative.

8.10   Mass increases with speed

We saw that it is impossible to speed up an object to reach c. This is because as we try to accelerate an object increasingly faster, it will offer increasingly greater resistance to our accelerating forces. Calculations from Relativity show that the mass of an object increases with its speed:

$$m = \frac{m_0}{\sqrt{1 - \frac{v^2}{c^2}}} \tag{8.4}$$

Suppose an object has mass 100 kg when it is stationary. When it is moving at a speed of 0.9c relative to an observer, this observer will find the mass of the object to be

$$\frac{100}{\sqrt{1 - 0.81}} = 229.4 \text{ kg} \cdot$$

If the object were to travel at the speed c, its mass would become infinity. What this means is that as the speed of the body approaches the speed of light its mass increases continuously without limit. As the mass increases, the momentum also increases without limit. Since force equals the rate of change of momentum, the amount of force needed to accelerate the body becomes greater and greater as the body approaches the speed of light. Since work energy equals the product of force and displacement, the work done in speeding up an object to speeds very close to that of light becomes incredibly enormous. All the energy available on the earth is insufficient to speed up even a tiny atom to the speed of light.[1]

But what about photons? Do they not travel at the speed of light? Yes, they do. But photons are not accelerated from rest to the speed of light. They always travel at the speed of light, not less, not more. A photon is emitted by an atom at the speed of light. A photon strikes an atom at the speed of light and is absorbed by the atom. A photon does not slow down before it is absorbed, and does not accelerate when it is emitted. But because the entire photon is absorbed or emitted, the process of absorption or emission is not instantaneous, and the Uncertainty Principle is obeyed, as discussed earlier.

8.11 Mass and energy

We saw in the previous section that when an observer finds the speed of an object increasing, the same observer will also find the mass of the object increasing. Where does the extra mass of the moving object come from? This seems to contradict the law of conservation of matter according to which the total mass of all the objects in the universe does not change.

Einstein modified this law. He showed that energy can be converted into mass, and vice versa. When the object is in motion it has kinetic energy. This kinetic energy is manifested as an increase of mass.

The kinetic energy of a moving
object is manifested as an increase
of its mass due to its motion

$\bigcirc$ $\qquad$ $\bigcirc \longrightarrow 10^8$ m/s

5 kg $\qquad$ 5.3 kg

---

[1]The equations of Special Relativity were derived for objects moving with constant velocity. But here and in Sec. 8.6 we have been dealing with accelerating objects. The validity of Special Relativity in these cases can be easily demonstrated. Let us imagine that we are a non-accelerating observer and we observe an object initially at rest. Then suppose the object accelerates to 5 m/s and maintains this speed for 5 seconds. Then it accelerates to 10 m/s and maintains this speed for 5 seconds, and so on. Now Special Relativity provides the equations for the relativistic mass, the relativistic length and the relativistic time interval of the moving object as long as the object has a constant velocity. So Special Relativity can be applied at each of the stages of the object's motion when it is not actually accelerating. So by breaking down the motion of the object into several stages we can treat it as a problem in Special Relativity.

With a little bit of mathematics, Einstein was able to show that when energy is converted to mass or when mass is converted to energy, there is a simple relationship between the mass and the corresponding energy. So when a mass of m kg is converted into energy E joules (or vice versa) the two quantities are related by the equation

$$E = mc^2 \qquad (8.5)$$

In a nuclear explosion, the total mass before the explosion is slightly higher than the total mass after the explosion. This loss of mass is converted into energy. Since c^2 is a large quantity, the amount of energy obtained is quite large.

Exercise 8.2. Using the formula $\epsilon = h\nu$ for the energy of a photon of frequency ν find the energy and hence the mass of a photon of light of wavelength 436 nm. A nanometer (nm) is 10^{-9} m. Take $h = 6.63 \times 10^{-34}$ and $c = 3.00 \times 10^8$ m/s.

8.12 Relativity and quantum theory

Special Relativity becomes significant only for speeds approaching the speed of light c, and so many introductory books on quantum theory do not devote much space to this subject. But as we have shown earlier, any measurement on an electron must be made through the electromagnetic field, and that happens through absorption and emission of photons. A photon always travels at the speed of light, and therefore from a conceptual point of view Relativity is very closely tied to quantum theory.

This close relationship between Relativity and quantum theory also has a mathematical component which we will introduce in the next chapter.

Exercise 8.3.

1. What should be the speed of an object if its measured mass m is twice its rest mass m_0? Express your answer as a multiple of c.

2. Find the relative speed of two space ships traveling towards each other at speeds 0.50c and 0.80c as measured by an observer on earth. Express your answer as a multiple of c.

3. Two Deuterium (H_1^2) nuclei (each containing one proton and one neutron — also called a deuteron) combine to form a Helium 4 (He_2^4) nucleus (containing two protons and two neutrons — also called an alpha particle). The mass of a Deuterium nucleus is $3.34358348 \times 10^{-27}$ kg and the mass of a Helium 4 nucleus is $6.64465675 \times 10^{-27}$ kg. Take $c = 3 \times 10^8$ m/s.

How much energy is released in this nuclear fusion reaction?

4. In laboratory X a free neutron is observed to decay into a proton, electron and neutrino 9.3 minutes after it was produced. How long was this neutron observed to be in existence before decaying when observed in laboratory Y that is moving with speed 0.9c relative to X?

8.13 Summary

The Special Theory of Relativity — or Special Relativity — rests on two principles: first, that the speed of light is the same no matter what be the speed of the source or the observer, and second, that no experiment can be performed inside a closed laboratory to determine how fast the laboratory is moving, as long as the laboratory is not accelerating.

These two principles lead to some remarkable conclusions. First, no object can be accelerated to travel at a speed greater than that of light. Secondly, the measured length, mass and flow of time on a moving object will depend on the relative speed between the object and the observer. This leads to the conclusion that temporal ordering of events is not absolute. If event A occurred first and was followed by event B as measured in one reference frame, then it is possible that as measured in a different reference frame event B would be earlier than event A.

Because the measured mass of an object depends on the relative speed between the object and the observer doing the measurement, the kinetic energy of an object is proportional to the increase in mass of the moving object. More generally, the relation between the energy E of an object and its mass m is given by the formula $E = mc^2$ where c is the speed of light.

Chapter 9

The Geometry of Space and Time

9.1 Space time

9.1.1 *World lines*

In the study of motion and velocity we use displacement-time graphs. Consider a straight road that runs west to east. Suppose a car is moving at a constant speed on this road. A stationary observer — whom we shall call Alice — could draw a graph of the position of the car at different times and obtain something like this:

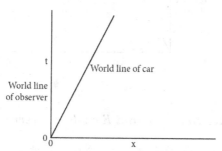

The horizontal line labeled x — called the space axis — represents the road along which the car travels from west to east. The vertical line labeled t — called the time axis — represents the flow of time.[1] The observer Alice sees the car at different positions on the road at different times. If the car is at a point labeled $x = 0$ at the starting moment when time $t = 0$ then the oblique line is the graph of the motion of the car. We call the oblique line the *world line* of the car. Alice herself is not in motion in space, but

---

[1]Displacement time graphs can also be drawn with time along the horizontal axis and displacement along the vertical axis. Either way the physics remains the same. See the first paragraph of Sec. 9.2.

her watch does show the flow of time. So she is traveling forwards not in space but in time. So the time axis represents Alice's world line.

The imaginary number i is defined by the equation $y^2 + 1 = 0$. We cannot get a real solution to this equation. The two mathematical solutions for y are written as i and $-i$. Other imaginary numbers can be defined by multiplying the unit imaginary number i by some real number, such as ai or $-bi$ where a and b are real numbers. One can also combine real and imaginary numbers by addition and subtraction to get *complex numbers* such as $5 + 3i$ and $4 - 7i$, etc. It is easy to see that $i^2 = -1$. A product of two complex numbers can be a real number. So if a and b are real numbers, then $(a + ib)$ and $(a - ib)$ are complex numbers, but the product $(a + ib)(a - ib)$ is a real number because it is equal to $a^2 + b^2$.

In the Special Theory of Relativity the time axis can be treated as a space axis if we multiply the time t by the speed of light c and the imaginary number i. We shall explain this strange rule in the following section.

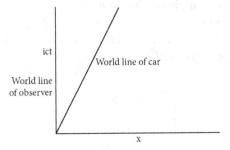

9.1.2 *Space-like, time-like and light-like intervals*

Suppose we have an interval of space Δx such as the distance between the earth and the sun (about 150 million km) and an interval of time Δt equal to 3 minutes. Clearly, it takes longer than 3 minutes (actually about 8 minutes) for light to reach the earth from the sun. So if an event occurred at 8:00 am US Central Time on the sun and another event occurred at 8:03 am US Central Time on the earth (on the same day) these two events are separated by a distance greater than the distance traveled by light in 3 minutes. We say these two events are separated by a *space-like interval*. So for this interval $\Delta x > c\Delta t$ implying that $\Delta x^2 > c^2 \Delta t^2$ and therefore $\Delta x^2 - c^2 \Delta t^2 = k^2$ where k is some real number. (A real number can be positive or negative, but the square of a real number is always positive.)

We could write this equation as

$$(\Delta x)^2 + (ic\Delta t)^2 = k^2 \tag{9.1}$$

This looks very similar to the equation for an interval in two-dimensional real space $(\Delta x)^2 + (\Delta y)^2 = \Delta r^2$. See the figure below:

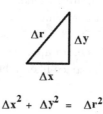

$$\Delta x^2 + \Delta y^2 = \Delta r^2$$

So Eq. (9.1) can be thought of as representing a different sort of interval, in which space and time are treated as equivalent.

Equation (9.1) is the equation for a space-like interval. Two events separated by a space-like interval are truly independent of each other. One event cannot affect the other. It would take longer for a message at light speed to travel from one place to the other than the time interval between these two events in the two places. So people who are present at one event cannot communicate the news of that event to the other place before the second event occurs.

Now, if we consider an event on the sun at 10:00 am US Central Time and another on the same day on earth at 11:00 am US Central Time the spatial and temporal intervals are now related by the inequality $\Delta x < c\Delta t$, because the time difference between the events is greater than the time taken for light to move from the location of one event to the location of the other. Such an interval is called a *time-like interval*. This would lead to the equation

$$(\Delta x)^2 + (ic\Delta t)^2 = -k^2 \tag{9.2}$$

where k is a real number. This is the equation for a time-like interval. Events that have a time-like interval may not always be independent, because it could be possible for a message to be sent from one event which arrives at the other place before the other event occurs.

It is also possible that $k = 0$, in which case the interval is called a *light-like* interval, because the distance between the events is equal to the distance traveled by light during that time interval. Are these events independent of each other? A message can in principle be sent from one event to the other at the speed of light, but since it takes some time — however small — for the message to be read and interpreted, events separated by an exact light-like interval are independent.

> **Exercise 9.1.** A distant star has two planets. A volcano erupts at 02:12:34 AM (hours:minutes:seconds) according to some interstellar standard time on one planet. A meteor hits the surface of the other planet at 11:56:43 AM the same day. If these planets are 1200 million kilometers apart, determine if the interval between the two events is space-like, light-like or time-like.

9.1.3   *Minkowski space*

In ordinary analytical geometry if the distance between two points is Δr and the differences in their coordinates are Δx and Δy then we have the Pythagorean equality $\Delta x^2 + \Delta y^2 = \Delta r^2$. A comparison of this equation with Eqs. (9.1) and (9.2) suggests that we can define a new pair of coordinates (x, ict) which span a coordinate system called *space-time*, also known as *Minkowski space*.

Here time is treated on a par with space by multiplying by the unit imaginary number i and a scaling factor c. You may have a foot long ruler which measures intervals in space. You may also have a clock which measures intervals in time. Thus the foot ruler measures real intervals and the clock measures imaginary intervals. So imaginary numbers are just as "real" as real numbers.

So Einstein's innovation made time and space equivalent. But this equivalence becomes significant only when we deal with objects traveling close to the speed of light, as we saw in the previous chapter. If the speed of light c were to become infinity then the equivalence between space and time would break down and the relativistic equations of the last chapter would revert to those of Newtonian physics. But the finite speed of light is a reality, and so space and time are interconnected in a real way. This interconnection allows us to represent microscopic processes involving one or more particles in a simple, elegant fashion.

9.2   Feynman diagrams

If time is indeed equivalent to space then the two axes — time and space — could be interchanged without violating the laws of physics. Interchanging space and time means rotating the coordinate axes through 90^0. Consider a physical process like the car moving along the road in the graphs shown above. This is obviously a physically observable event. If we switch the axes we would get another oblique straight line graph. This is qualitatively the same as before. Switching the axes does not appear to have yielded anything new. But there are other cases which are more interesting. Let us see some examples.

Suppose Alice is observing an electron moving away from her, which then emits a photon, and because momentum is conserved, the electron recoils in the opposite direction towards Alice. The electron world line is shown as a thick line, and the photon world line is shown as a thin line.[2]

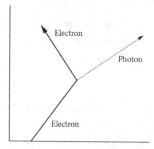

A diagram such as the one shown above is called a *Feynman diagram* or a *Feynman graph*. Such diagrams are extremely helpful for visualizing the process and for performing calculations to determine numerical quantities associated with the process.

A particle which has the same mass but the opposite charge as the electron is called the positron. If an electron ran into a positron, the two particles would annihilate each other and generate two photons. The total momentum and energy will be conserved. We can represent this process by a Feynman diagram. In the figure below, an electron travels in the positive x direction and a positron in the negative x direction. After their mutual annihilation, two photons are created which travel in opposite directions:

---

[2]It is common to depict photon world lines as wavy, and those of electrons and other particles as straight. But there are two drawbacks with this convention. The first is that the waviness of the photon line could suggest that a photon has more wave properties than an electron, which is not true. Secondly, a wavy world line could convey the erroneous notion that the photon executes an undulating motion in space and time, which is not the case. In this book we will freely depict photon world lines as straight lines or as wavy lines and choose the latter only when the waviness illustrates a point. So we did that for the Compton effect to underscore the fact that light — conventionally thought of as a pure wave — can exchange momentum with an electron.

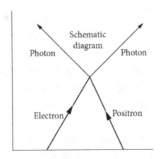

Actually the above diagram is not quite accurate, because an electron (or positron) interacting with a photon actually requires two vertices, with two electron (positron) lines meeting a single photon line. We shall explain later why this must be so. But the schematic diagram shown above is sufficient to illustrate the concept of world lines.

9.3   Arrow of time

Time flows relentlessly forwards. We do not grow young, ripe fruits do not turn green, the pieces of a broken glass vessel do not unite to form the complete utensil, and so on. The reason for this is that all things are made up of a very large number of component molecules. These molecules obey statistical mechanics, and as we saw in Ch. 3 their motion tends towards greater disorder than order, a fact that is expressed as the Second Law of Thermodynamics. Equivalently, we could say that the energy available to do work is decreasing, or that there is a progressive loss of information. The scientific term for this disorder or lack of energy or information is *entropy*. So the natural tendency of all material systems is towards increasing entropy. The entropy of the universe is constantly increasing. And this is what we subjectively perceive as the flow of time. And this is why the arrow of time points forward, never backwards. It is pure fiction to imagine a travel to the past, though a one way travel to the future is permitted by Relativity. Actually we are constantly traveling one way to the future, but Relativity shows how this could be accelerated if we have the resources to burn up a lot of energy.

An hour glass cannot run backwards, because the sand cannot flow upwards against gravity.[3] Pendulum clocks depend on the potential energy stored in a spring that needs to be wound periodically. A spring cannot

---

[3]This process is irreversible. As the grains of sand collide with each other and with the walls of the glass container they lose kinetic energy which is converted to heat. So the

wind itself spontaneously. Battery operated watches will stop working when the battery runs down. Both the spring and the battery are examples of processes where the amount of available energy decreases with time.

In classical mechanics we distinguish between elastic and inelastic collisions. In an elastic collision the total kinetic energy after the collision between the objects is equal to the total kinetic energy before the collision. In an inelastic collision some of the kinetic energy of the colliding objects is converted into heat energy and therefore there is a lessening of kinetic energy. No real collision is perfectly elastic. Some of the kinetic energy of the bodily motion of the objects is converted into kinetic and potential energies of the individual molecules of the objects — which is heat — and therefore every collision of ordinary objects is an irreversible process.

The arrow of time is most evident in our subjective experience of memory. Memory works on irreversibility. A reversible process does not leave a memory. In the real world only microscopic processes are truly reversible. When two electrons collide with one another such collisions are elastic, because no part of the kinetic energy is converted into heat, for the concept of heat is meaningless when we are considering individual microscopic particles. Thus every collision between pairs of such elementary particles is time reversible.[4] We could play the tape backwards and the process would be physically possible. So a particle can actually move backwards in time. But a particle that moves backward in time does not take with it any memory of the present, because a single particle cannot carry information about events. So while it is possible for individual particles to move backwards in time, they cannot possibly do anything to the past which will alter the present.

This distinction between single particles and large collections of particles as regards time travel is related to the Second Law of Thermodynamics which is a statistical law deriving from the very large number of molecules in any macroscopic body.

9.3.1 *Time reversal and Feynman diagrams*

Negatively charged electrons had been discovered by J. J. Thomson. Later, when Dirac applied the theory of Relativity to the quantum mechanics of electrons, he discovered that there had to be another particle, identical

---

gravitational potential energy is converted to kinetic energy as the sand falls through the narrow passage and this increased kinetic energy is lost as heat energy.

[4]We are excluding collisions where particles are destroyed or new particles are created.

in mass with the electron, but exactly opposite in charge, and therefore positively charged. This particle was called the positron, and it was subsequently discovered experimentally. A positron is called an *antiparticle* of an electron. We will have more to say about these matters in a later chapter.

Positrons are not very stable, because they are attracted by the negatively charged electrons which are so abundant in nature, and when these two particles come very close to each other, they literally fall into each other and annihilate each other. In that process, energy is released in the form of photons. There has to be a minimum of two photons in this process. We reproduce below the Feynman diagram for this process:

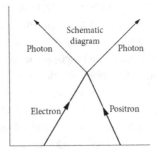

Feynman had this brilliant idea that the positron could be looked upon as an electron traveling *backwards in time*. In the diagram above, we see an electron and a positron approach each other — both traveling forwards in time — and then annihilate each other to produce two photons — both traveling forwards in time. Feynman suggested that we think of the electron — positron line as a single world line of a single electron — moving forwards in time, emitting two photons and then moving backwards in time. A negative particle traveling backwards in time would appear as a positive particle traveling forwards in time.

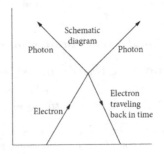

Suppose we have a positive charge + Q coulombs fixed to some point in space so it cannot move. Next, we picture an electron of charge −e in the vicinity of this positive charge. If the electron is released and allowed to move freely it would rush towards the positive charge. As we take a movie clip of this process, the electron will be seen accelerating towards the positive charge.

Next, we play the move clip backwards. Now we see the electron moving away from the positive charge, and decelerating as it moves further and further away. This motion would correspond to that of a positive charge being repelled by the charge + Q. Thus an electron traveling backwards in time is indistinguishable from a positron traveling forwards in time. (Of course, in this scenario our interest is only in the nature and motion of the electron, not the stationary positive charge. So we should clarify that our *system* consists only of the electron or positron, and that the fixed positive charge is an *external* source of an electric field. In other words, we are not interested in the question of what would happen to the fixed positive charge if that were also subject to a time reversal.)

What about a photon? This particle has no charge, and so it is its own antiparticle. *So a photon can be thought of as moving either forwards or backwards in time.* So we could just as correctly draw the diagram thus:

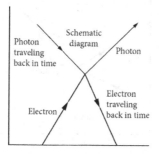

Photons also have this unique property that they travel at the speed of light c. And so if we were to do something so weird as to try to observe the rate of time flow on board this photon, we would find that time does not flow at all for the photon, since a time interval T_0 on the photon measured by us as T would be

$$T_0 = T\sqrt{1 - \frac{v^2}{c^2}}$$

which would be 0 for a photon with $v = c$.

The backward flow of time for processes involving single particles has some implications for colliding particles. We saw that an electron and a positron can annihilate each other and produce two photons. Now, if this process is time reversed, we would observe two photons colliding and producing an electron and a positron. This process does take place, and is called electron positron pair production.

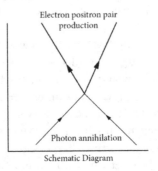

Schematic Diagram

Thus we see that by inverting the Feynman diagram we obtain a different physical process. According to Special Relativity both sets of coordinate axes are physically valid. Now suppose we were to rotate the axes through a right angle, so that the space and the time axes are interchanged. Then the process of electron positron mutual annihilation would look like this:

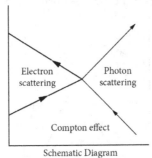

Schematic Diagram

This is the Feynman graph of a Compton Effect (Ch. 6).

So we can enunciate a rule for quantum processes thus: If a particular Feynman diagram represents a possible physical process, then all the processes that can be imagined through a rotation, inversion or reflection of the Feynman diagram are also physically possible. One can also change the directions of the arrows and reinterpret the particles corresponding to those arrowed lines.

9.3.2 *Information carried by many photons*

We must stress once again that a single photon cannot carry any useful information forwards or backwards in time. A single photon coming to the earth from a distant galaxy can tell us nothing about how far away that galaxy is, or when the photon was emitted by the galaxy. We need billions of photons diverging from the distant object that can give us details of that object.

It is only because billions of photons coming from the moon enter our eyes and form images on our retinas that we are able to perceive that the photons came from a far away object. In order for an image to be formed, the photons must be transported by waves that are capable of reflection and refraction. We saw in Sec. 7.4 that the wave nature of light can be demonstrated only after a very large number of photons have created an interference pattern on the screen. Thus, for light to be manifested as a wave, there must be billions of photons involved.

Hubble showed that it is possible to tell how far away a distant galaxy is by observing how fast it is moving away from us. In order to understand how this is done, let us recall that every star is made up of burning gases. When hydrogen burns, it emits light of characteristic wavelengths (cf. Ch. 7). A spectrometer can measure these wavelengths, and so knowing the pattern of wavelengths — called a *spectrum* — emitted by hydrogen, we can tell if a star has hydrogen. If such a star were moving away from the earth at a high speed, the light reaching us from the star would undergo a stretching or an increase of its wavelength due to a phenomenon called the Doppler effect. By measuring the amount by which the wavelength has increased it is possible to calculate the speed of the star. Hubble measured the speeds of galaxies whose distances from the earth were known by other means, and found that the further away the galaxy was from the earth, the greater was the stretching of the wavelength and therefore the faster it was moving.[5] Such a stretching is called the *red shift* (because red light has the longest wavelength among the colors of visible light). Hubble's discovery led to the realization that the universe is expanding. The expansion of the universe means that the very fabric of space itself is expanding, and this expansion carries galaxies away from each other. This also means that the further

---

[5]Cepheid variables are pulsating stars whose true luminosity (rate of emission of energy) is related to their pulsation rates. By observing their pulsation rates we can calculate their true luminosity. Then by comparing their true luminosity with their observed luminosity we can gauge their distance from us. Cepheid variables have been identified in our galaxy as well as in other galaxies.

away a galaxy is from the earth, the faster it is receding from the earth. Since the red shift helps us to measure the speed of the galaxy, it also helps us determine how far away is the galaxy. Thus, measuring the wavelength helps us calculate the distance of the galaxy from the earth. But wavelength is a wave property, and in order to measure wavelength we need a very large number of photons.[6]

Thus, in order to obtain any information about a distant object, we need an accumulation of photons collected at different points of our detecting apparatus. A single photon cannot communicate that sort of information. And while a single particle such as a photon or an electron can travel backwards in time, it cannot carry back any useful information with it.

Exercise 9.2. A solar flare erupted on the sun at 10:07 AM New Orleans Time on the 15th of March. A car crashed into a telephone pole on Claiborne Avenue at 10:10 AM the same day. The driver later learned about the solar flare, and claimed compensation from the car company, on the grounds that the radiation from the solar flare interfered with the computerized steering of his car. The distance of the sun from the earth is 150,000,000 km.

(a) Is the interval between the events (flare on the sun and accident in New Orleans) space-like or time-like? Show your calculations.

(b) Could one event have had an effect upon the other? Explain.

9.4 Summary

Einstein's Special Theory of Relativity brought space and time together. One is no longer independent of the other. But space and time are not really equivalent. An interval in time is not the same as an interval in space. But Einstein showed that the two intervals could be made mathematically equivalent by the factor i which is called the unit imaginary number, defined by the equation $i^2 = -1$. So one can create a four-dimensional space which we call *space time* or Minkowski space, in which one coordinate is imaginary, representing the flow of time, and the other three coordinates

---

[6]Wavelength can be measured by interference, and as we saw earlier in Ch. 7 a single photon will not produce interference patterns.

are real, representing space.[7] Such a coordinate system represents a frame of reference, and every observer has their own frame of reference. That is, every observer can record the positions of all the objects he or she observes on such a graph. A two-dimensional representation is commonly shown with time along the vertical axis and one dimension of space along the horizontal axis. So if there is a stationary object such as a tree, that object will have a fixed position in space but will be moving in time. This "path" of the tree is a straight line parallel to the time axis and perpendicular to the space axis. Such a path is called a *world line*. A moving object will have a world line that is at an angle to the time axis.

Any two events are separated in both space and time. The separation between them is called the interval. The magnitude and the sign of the interval are determined by the speed of light. If the two events occur at two times that are very close in time, but the places where they occur are exceedingly far away, then we say that the interval is space-like. For a space-like interval, the time taken for light to go from one event to the other is greater than the time that elapses between the events. If two events occur in places close to each other, and the time difference between them is sufficiently large, then the interval is time-like. If the events are separated by a time difference which is exactly equal to the time taken for light to travel from the place of one event to the place of the other, then the interval is called light-like.

A diagram that shows the world lines of two or more particles that interact with each other is called a Feynman diagram or a Feynman graph. Feynman graphs help us to provide a graphical understanding of the process by which one particle interacts with the other. One important feature of Feynman graphs is that a graph may be rotated or inverted and the resulting world lines would have different physical interpretations, but the remarkable thing is that all such interpretations would be physically meaningful.

Feynman graphs offer multiple interpretations. For example, if an electron meets a positron they would annihilate each other and produce two photons. But the positron world line can be thought of as an electron world line traveling backwards in time. So Feynman diagrams imply that single particles can travel backwards in time.

Whereas individual particles can travel backwards in time, it is a different matter with large collections of particles such as space ships and human beings. Such collections cannot travel back in time. And so time travel is

---

[7]Today, the imaginary number is eliminated from the mathematics of Relativity using what is called a metric tensor.

impossible for humans (or for animals or for robots). In general, useful information such as a photo of an event can be carried only by a very large number of particles. A single particle cannot carry that sort of information forwards or backwards in time.

Chapter 10

The Heart of Quantum Theory

10.1 How does one study the quantum?

Classical electromagnetism deals with the laws of electricity and magnetism as they were known prior to the twentieth century. According to these laws charges receive or emit electromagnetic energy in continuous streams. But Planck showed at the beginning of the twentieth century that energy can be absorbed or emitted only in discrete quanta of magnitude $h\nu$. This idea is the genesis of quantum theory and remains one of its central tenets.

Einstein took Planck's hypothesis one step further and showed that the reason why light is absorbed or emitted in packets or quanta is that light *exists* as packets or quanta. And these quanta of light are particles in their own right, with energy, mass, momentum, and even angular momentum. They collide with electrons in the phenomenon called the Compton Effect. Light propagates as a wave but is detected as a particle. De Broglie removed the basic distinction between light and matter when he showed that the so-called material particles such as electrons also have a wavelength. They too propagate as waves but are detected as particles. The study of the energy, motion, angular momentum and other properties of a physical system is called *mechanics*. Classical mechanics is the study of large objects such as meteors, footballs, automobiles, the moon, the stars, and other such entities containing billions of atoms. Quantum mechanics is the study of microscopic objects such as photons, electrons, protons, neutrons, and small collections of particles such as atoms and molecules.

We now come to the heart of the difference between classical and quantum mechanics. In any problem in classical mechanics we can concentrate on one aspect or one facet of the object and ignore the others. As an example, suppose we are interested solely in studying the forces that lift up a flying airplane. For most purposes it is sufficient to study the air flow above and below the wings to understand how the aircraft overcomes gravity and remains aloft. The remainder of the physical properties of the airplane — such as the pressure or the temperature in the interior — is largely irrelevant for the investigation of the lift.

By contrast, there are no parts or components of a fundamental particle such as an electron or a photon. When we observe or detect an electron we do not detect its top part or its side.[1] We detect the entire particle. Likewise for a photon.

Quantum theory as it is understood today states that in order to make any sort of measurement on a particle, the particle must be absorbed first, and then re-emitted. And the absorption and re-emission involve the whole particle, not any one portion of it. Let us illustrate this in the case of the scattering of an electron in the Compton Effect. The schematic Feynman diagram illustrating this phenomenon is the following:

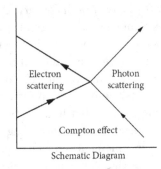

Schematic Diagram

However, this is not the correct Feynman diagram. A correct Feynman diagram cannot have a vertex with two electron lines and two photon lines meeting at a point. It must have two electron lines and a photon line meeting at a point. So there are two such vertices in the complete Feynman diagram:

---

[1]One does make a *distinction* between the electron's charge, mass and spin, but it is impossible to separate them.

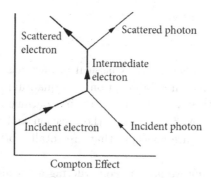

Compton Effect

There are five *distinct* particles in this diagram: the incident electron, the incident photon, the intermediate electron, the scattered electron, and the scattered photon. The incident photon disappears at the first (lower) vertex, and the "scattered" photon appears at the second (upper) vertex. It is common in quantum theory to use the terms "annihilated" and "created" for the disappearing and the appearing photons. These terms serve to indicate that the incident photon comes to the end of its career when it is absorbed by the electron, and the "scattered" photon begins its own career when it is emitted by the electron. So they are two different particles.

What is perhaps really surprising is that the incident electron also ends its career when it absorbs a photon. It is annihilated the moment it absorbs the photon. The intermediate electron is a new particle that exists only for a short time before it is annihilated and a new electron is created which appears as the "scattered" electron.

It is easier to think of the electron as a continuously existing particle because it has charge, and charge cannot be destroyed according to Maxwell's theory of electromagnetism. But quantum theory permits a charge to be destroyed and recreated immediately.

In the historical development of quantum mechanics particles such as electrons were treated as though they existed indefinitely, even as they interacted with other charges and with photons. The notion of electrons being created and destroyed as they interact with other particles is a later development, and this quantum mechanical approach was called *second quantization* or *quantum field theory.* But it has now been recognized that this so-called field theoretical approach expresses a better understanding of the behavior of fundamental particles such as electrons and photons.

10.2   Fields and states

In classical physics objects *exist*. Classical physics has absolutely no way of describing the emission or absorption of a quantum of electromagnetic energy. Maxwell's theory could only describe continuous emissions or absorptions of energy. And, of course, the creation and annihilation of an electron or a positron are concepts that are totally foreign to Newtonian physics.

Thus the first thing we need in constructing an accurate description of quantum theory is a mathematical base that incorporates the notions of absorption and emission. Such a mathematical base is the foundation of *quantum mechanics.*

Historically, quantum theory arose with the discovery of light quanta or photons. It is therefore appropriate to commence our study of quantum mechanics with photons.

First of all we need a backdrop against which a photon can appear. And this backdrop is called a *field.* And the field in which a photon can appear is called an electromagnetic field, which is the same term used in classical electromagnetism, except that it is now understood differently. If a photon appears in the field, we say that the state of the field has changed. If two photons appear, that is a different state than the state having a single photon.

So, the mathematical description of quantum theory begins with the notion of a *state*. The concept of a state is the basis of all quantum mechanics.

Next, we need to describe the appearance and the disappearance of a photon. The sudden appearance or emission of a photon has been given the religious sounding term *creation*. And the disappearance or absorption of a photon is called *annihilation.*

Since absorption and emission are *processes*, they are different from states. A process is described by an entity called an *operator*. Thus we talk about a creation operator and an annihilation operator.

The essence of quantum mechanics lies in the relationship between the states and the operators. And this relationship constitutes the language of quantum mechanics.

In this chapter we shall develop this new language in the context of photons. Of course, this language is also applicable to electrons, protons, neutrons, etc.

10.3   Complex numbers in quantum mechanics

Ultimately, all mathematical entities or symbols have a numerical component. They are expressed in terms of numbers. Everything that is observable and measurable must be expressed by a real number. Length, speed, mass and energy cannot be imaginary. However, the intermediate stages in the description of a physical process need not be restricted to real numbers.

In the case of quantum theory, it turns out that operators are best described by both real and imaginary numbers. In other words, operators in general require complex numbers for their description. We shall offer a simple qualitative explanation for why this must be so.

In classical physics one can separate the spatial and temporal aspects of any process. For example, in order to measure velocity we make two separate measurements — a measurement in space for the displacement, and a measurement in time for the time interval, and we divide the displacement by the time interval in order to obtain velocity.

In quantum physics we cannot separate space from time. Any process of measurement brings together space and time in an inextricable union. According to the Uncertainty Principle, a measurement of momentum invariably leads to an uncertainty in our measured value of position, and vice versa. Since a measurement of momentum is a measurement of motion, which necessarily involves time at least implicitly, and a measurement of position requires a measurement of space, we see that every quantum measurement links space and time in an intimate relationship.

Now, space and time are related at the fundamental level by the imaginary number i, as we saw in the preceding chapter. So now we see that it is impossible to do quantum mechanics without imaginary numbers.[2] And so it should come as no surprise that complex numbers are necessary to give a complete description of quantum processes. But it may be that in some situations the imaginary part of the complex number becomes zero, and so the number is purely real.

Let us describe the process of absorption by a mathematical symbol a. Right now we do not know anything about a except that it describes the absorption or annihilation of a photon. Of course, the letter a is just a symbol or a name for a mathematical entity that we will use to provide

---

[2]It is in principle possible to write the equations of Relativity and Quantum Theory without imaginary numbers, but such a procedure will amount to a camouflaging of the imaginary numbers. As an example, we can do arithmetic without using the number 7. So we would write the number 37 as $35 + 2$, 77 as $80 - 3$, etc.

a mathematical description of the absorption or annihilation of a photon. Since this entity describes an action it is called an *operator*. And since it describes the action of annihilation it is called the *annihilation operator*. And we expect that the imaginary number i figures in this operator in some way. The operator representing emission or creation is written as $a^\dagger$ and is called the *creation operator*. So we still do not know a great deal about these two operators other than that they change the state of the field and that we probably require imaginary numbers in addition to real numbers to express them mathematically.

10.4 States and operators

Let us return to the black body cavity. Let us consider these two distinct processes:

Process I: Wall A emits a photon
Process II: Wall B absorbs a photon.

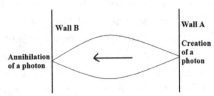

In this process a photon is first created and then annihilated. We now represent this process mathematically.

Suppose there were no photons in the cavity prior to the creation of this photon. We express this state of the cavity by the symbol $|0\rangle$. Such a symbol is called a *state function*, or a *state vector*, because it expresses the *state* of the cavity.[3] This particular symbol with a zero means that this function or vector represents a state with no photons.

The *state* — (the word is related to the adjective *static*) — does not describe a process. A process is described by an *operator* (Latin for someone or something that performs work). As stated earlier, the process of emission of a photon is described by a *creation operator* written as $a^\dagger$. An operator works on a state. So the full description of the emission of a photon is expressed in quantum mechanical symbolism as $a^\dagger|0\rangle$. And since the emission of a photon is a time dependent process, we expect that the mathematical description of this process should contain the imaginary number i.

---

[3]The word *function* is an algebraic term, and the word *vector* is a geometric term, but they express the same physical reality.

The result of this process is that the cavity now contains one photon of light. But the photon is subsequently absorbed by wall B and so the cavity returns to its empty state. The process of absorption is the exact time reverse of emission. In quantum mechanics we represent the process of absorption by means of an absorption or annihilation operator which we write as a. So the entire process can be written as $aa^\dagger|0\rangle$. And since we now have 0 photons in the cavity, we are back to the state $|0\rangle$. So the overall effect of the two operators $a^\dagger$ and a acting one after another is that there is no net change to this particular state, i.e. $aa^\dagger|0\rangle = |0\rangle$.

10.5 Physical meaning of symbols

10.5.1 *Creation and annihilation of photons*

How would we know that $aa^\dagger|0\rangle$ is identical with $|0\rangle$? We can check the state $aa^\dagger|0\rangle$ to see if it has no photon. This has to be done physically. And the mathematical way we express the process of checking this is to write the symbol $\langle 0|$. This symbol means we are checking the number of photons to see if indeed there is no photon. But we are not asking the question: How many photons are there? Rather, we are asking: "Is it true that there is no photon in the cavity?" And so the entire physical process is expressed by the mathematical symbols

$$\langle 0|aa^\dagger|0\rangle$$

In quantum mechanics we read these symbols from right to left. The earliest stage is at the extreme right, and the latest on the extreme left. So this symbol is shorthand for saying that initially there were no photons in the cavity ($|0\rangle$), then a photon was created ($a^\dagger$), then a photon was annihilated (a), and finally we checked that there were no photons in the cavity ($\langle 0|$).

The expression $\langle 0|aa^\dagger|0\rangle$ may be read as a *report* of the entire process. It is a communication of *information*. While the symbolic expression does have a narrative appearance, it also has a numerical value. The numerical value of the report is a number, in general complex (since both the creation and the annihilation operators contain imaginary numbers). If the process is certain or inevitable, we give the report a value of 1. If the process is impossible, we give it a value of 0. If there is some possibility of it occurring even though it is not entirely certain, we give it a number — which we shall call ψ — whose *absolute value* is intermediate between 0 and 1. The absolute value of a complex number $a + ib$ (where a and b

are real) is defined as $|a + ib| = \sqrt{(a + ib)(a - ib)} = \sqrt{a^2 + b^2}$, keeping in mind that the square root symbol always indicates a positive number. If $\psi = a + ib$ where a and b are real, then we define the *complex conjugate* of ψ as the number $\psi^* = a - ib$. So $|\psi| = \sqrt{\psi\psi^*}$.

Exercise 10.1.

1. $|3.00 + 5.00i| = \sqrt{3^2 + 5^2} = \sqrt{9 + 25} = \sqrt{34.0} = 5.83$. Find the values of
(a) $|4.00+3.00i|$ (b) $|6.00+8.00i|$ (c) $|2.00+2.00i|$ (d) $|0.05+0.12i|$

2. Given $\psi_1 = 4.0 + 5.0i$ and $\psi_2 = 3.0 - 2.0i$ find
(a) $\psi_1\psi_2$ (b) $\psi_2\psi_1$ (c) $|\psi_1|$ (d) $|\psi_2|$ (e) $\psi_2^*\psi_1$ (f) $\psi_1^*\psi_2$

Thus, the report of the entire process is a (complex) number. But the intermediate stages of the process are not numbers. The initial state $|0\rangle$ is not a number but a state function or a state vector. The process of the emission of a photon is not represented by a number but by an operator $(a^\dagger)$ which is not a number. When the creation operator acts on the state with no photons, it changes the state into a state with 1 photon. This process is written as

$$a^\dagger|0\rangle = |1\rangle \tag{10.1}$$

An annihilation operator a performs the reverse operation as the creation operator. So if a were to operate on $|1\rangle$ it would give us $|0\rangle$:

$$a|1\rangle = |0\rangle \tag{10.2}$$

What do Eqs. (10.1) and (10.2) represent physically? The precise meaning will depend on the exact physical conditions. As an example, they could represent the interaction between an atom and a photon. The first expresses an atom emitting a photon and the second represents an atom absorbing a photon. So quantum mechanics offers us an elegant mathematical language for expressing the fundamental tenet of quantum theory, that energy is absorbed or emitted not continuously, but in units or quanta.

If we knew the actual physical conditions of the system in which a process occurs, then we could calculate the numerical value of the report of this process, a value which in general is a complex number whose absolute value lies between 0 and 1. What is the purpose of calculating this number? It turns out that every prediction that we can make in a particular situation depends on the calculation of this number for that situation.

> **Exercise 10.2.** Find the values of the following numbers:
> (A) $\langle 1|aa^\dagger a|1\rangle$ (B) $\langle 0|a^\dagger aa^\dagger|0\rangle$ (C) $\langle 1|a^\dagger aa^\dagger|0\rangle$ (D) $\langle 0|a^\dagger aa^\dagger aa^\dagger|0\rangle$

10.5.2 *Propagation of a photon*

Suppose now the cavity is divided into two chambers by a wall with two openings labeled I and II. Next, suppose a photon is emitted at A and absorbed at B. How do we express this process quantum mechanically? We first write an expression for the creation of a photon in the right-hand chamber: $a_r^\dagger$. So we now have one photon in the right-hand chamber, and we can express this state as

$$a_r^\dagger|0\rangle_r$$

where the subscript r signifies that we are considering the right-hand chamber.

Now, the photon can pass through either of the two openings I or II. Let us see what happens if the photon passes through I. The passage of the photon can be thought of as two consecutive actions — entering the opening from the right chamber, and exiting the opening into the left. Each of these is a separate action and needs a separate operator to describe the process. We represent the photon's entering the opening I by the operator $U(I)$. So the process so far reads:

$$U(I)a_r^\dagger|0\rangle_r$$

In the next step, the photon exits the opening I. Since entering the opening was described by the operator $U(I)$, we shall represent the exit of the photon from the opening by the operator $U^\dagger(I)$, where the dagger symbol indicates an operator representing a reverse process, similar to the creation and annihilation operators. So the process now reads:

$$U^\dagger(I)U(I)a_r^\dagger|0\rangle_r$$

Finally, the photon is absorbed in the left-hand chamber and disappears. This step is represented by the operator a_ℓ. The state is then checked and no photon is found in the left chamber. This final step is represented by the checking symbol $\langle 0|_\ell$. The report of the entire process is given by:

$$\langle 0|_\ell a_\ell U^\dagger(I)U(I)a_r^\dagger|0\rangle_r$$

This report is a number, but we cannot give it the value 1. The reason is that this is not the only possible way for an electron originating in the right chamber to be absorbed in the left chamber. There is the possibility that the photon might have gone through opening II. So let us say

$$\langle 0|_\ell a_\ell U^\dagger(I) U(I) a_r^\dagger |0\rangle_r = \psi_1$$

where ψ_1 is a complex number and $|\psi_1|$ lies between 0 and 1.

Next, we consider the possibility that the photon went through the opening II. We would get a similar number for this path:

$$\langle 0|_\ell a_\ell U^\dagger(II) U(II) a_r^\dagger |0\rangle_r = \psi_2$$

where ψ_2 is another complex number and $|\psi_2|$ also lies between 0 and 1.

10.5.3 *Probability amplitudes*

The numbers ψ_1 and ψ_2 are called *probability amplitudes* or simply *amplitudes*. So ψ_1 is the amplitude for the photon to go through I and ψ_2 the amplitude for the photon to go through II.

We saw above that a photon (or an electron or a proton, etc.) propagates as a wave but is detected as a particle. While it is passing through the wall separating the two chambers *the photon is not detected*. So we must consider the dynamics of the photon's movement from one chamber to the other to be the dynamics of a wave. Now one of the properties of a wave is interference. So the photon wave — or light wave — undergoes interference as it moves between the two chambers. So there is interference between the wave that flows through opening I and the one that flows through opening II. In quantum theory, interference is expressed by adding the amplitudes. So the complete description of the photon being emitted at wall A and absorbed at wall B is given by

$$\psi_1 + \psi_2 = \langle 0|_\ell a_\ell U^\dagger(I) U(I) a_r^\dagger |0\rangle_r + \langle 0|_\ell a_\ell U^\dagger(II) U(II) a_r^\dagger |0\rangle_r$$

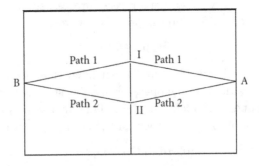

If we assume that a photon emitted at wall A *must* cross the partition and be absorbed at wall B, then the above equation tells the whole story.

This means that $\psi_1 + \psi_2 = 1$. We notice that the expression for each amplitude summarizes the path. The expression begins on the right with the state function and ends on the left with the state function in reverse form, signifying closure of the process through a checking. In between we have a string of operators.

10.5.4 *Addition of paths*

Thus each amplitude is a product of possibly several operators between two state functions at the two ends. This amplitude is a complex number. And we add up the amplitudes for all the possible paths.

If we know for certain that the process is bound to happen — i.e. a photon emitted at A is bound to cross the partition through either I or II and be absorbed at B, then the sum of the amplitudes is 1. But if the process is not one hundred percent certain — say there is a possibility that the photon could be absorbed by the partition itself, then $\psi_1 + \psi_2$ will not be equal to 1, but $|\psi_1 + \psi_2|$ would be less than 1. In this case the probability of this entire process occurring is given by the square of the absolute value of the amplitude:

$$\text{Probabiility of I or II occurring} = |\psi_1 + \psi_2|^2 \qquad (10.3)$$

So the rule is as follows:

1. Multiply all the operators representing a particular path, making sure that the order in which the operators are placed — right to left — follows the path correctly.

2. Multiply this operator product on the right by the initial state and on the left by the final state (written in reverse form). This product will be a complex number, and is called the amplitude for this path.

3. Add up the amplitudes for all the paths.

4. The square of the absolute value of this sum of the amplitudes is the probability for the entire process to occur. This could be anywhere between 0 and 1.

10.6 Classical and quantum probabilities

Suppose we were to close up one of the two openings, say opening II. Then the photon would be able to go through I only. The amplitude for this process is simply ψ_1. The probability for this process to occur is given by $|\psi_1|^2 = \psi_1^* \psi_1$.

If now we close up I and keep II open, the probability for the photon to go through this opening is $|\psi_2|^2 = \psi_2^* \psi_2$.

Now we come to one of the biggest differences between classical and modern physics.

Probability for the photon to go through I and NOT through II = $|\psi_1|^2$.

Probability for the photon to go through II and NOT through I = $|\psi_2|^2$.

If now both I and II are open, *according to classical physics*, the probability that the photon goes through either I or II is simply the sum of the probabilities:

$$|\psi_1|^2 + |\psi_2|^2 = P \text{ (either I or II)} = P_c.$$

But quantum physics makes a very different prediction. If both I and II are open, the probability that the photon goes through these openings is

$$|\psi_1 + \psi_2|^2 = P \text{ (both I and II)} = P_q.$$
$$P_q = |\psi_1|^2 + |\psi_2|^2 + \psi_1^*\psi_2 + \psi_1\psi_2^* = P_c + \psi_1^*\psi_2 + \psi_1\psi_2^*.$$

According to quantum theory the photon is detected as a particle, but it propagates as a wave. And so when both I and II are open the photon travels through these openings and there is *interference* between these two portions of the wave. This interference term is the difference between the classical and quantum probabilities, which equals $\psi_1^*\psi_2 + \psi_1\psi_2^*$. This term can be positive or negative.

Exercise 10.3. Prove that $\psi_1^*\psi_2 + \psi_1\psi_2^*$ must be a real number.

10.6.1 *Constructive and destructive interference*

The highest positive value for the expression $\psi_1^*\psi_2 + \psi_1\psi_2^*$ occurs when $\psi_1 = \psi_2$, in which case this expression has the value $2|\psi_1|^2$ (or $2|\psi_2|^2$). So then we would have $P_q = 4|\psi_1|^2$ and $P_c = 2|\psi_1|^2$ so that the quantum mechanical probability P_q is twice the classical probability P_c. This is constructive interference.

When $\psi_1 = -\psi_2$ the expression $\psi_1^*\psi_2 + \psi_1\psi_2^*$ has its lowest negative value equal to $-2|\psi_1|^2 = -2|\psi_2|^2$. In this case $\psi_1 + \psi_2 = 0$ and $P_q = 0$. But P_c remains the same at $2|\psi_1|^2$. So if we think of the photon as a classical particle, it has a probability of going through one or the other opening, but as a quantum mechanical wave particle, the photon cannot cross over even when both paths are open.

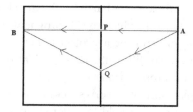

The above diagram illustrates how the quantum mechanical probability could become zero. Earlier we considered the general case of a photon created in the right chamber and absorbed in the left chamber. This time we pay attention to the specific spot on the wall of the left chamber where the photon was absorbed or detected as well as the specific spot on the right wall where the photon was emitted or created. Let us consider a photon created at A. A detector is placed at B to observe this photon. If we think of the photon as a classical particle, this photon has some probability of going through P and then to B, and another probability of going through Q and then to B. So according to classical physics the probability that the photon is detected at B is simply the sum of these two probabilities.

But this classical probability is incorrect, because it was calculated by assuming that the photon *propagated as a particle*, which is wrong. A photon, like any other particle, propagates as a wave, but is detected as a particle. The correct probability is the quantum mechanical probability. And this probability is calculated by considering the propagation of the photon as a wave. Because it is a wave, it undergoes interference between the two paths. One wave travels from A to P to B, and another wave travels from A to Q to B. Suppose the difference between these two paths is exactly an odd multiple of half a wavelength:

$$AQB - APB = (2n + 1)\frac{\lambda}{2}$$

where n is any whole number 0, 1, 2, 3, etc. Then there would be destructive interference between the two waves at the point B, with the result that the intensity at B would be zero. This means that there is zero probability of finding the photon at B. This is destructive interference.

10.7 Summary

The mathematics of quantum theory is called quantum mechanics. Every interaction involving a particle is described in quantum mechanics as an absorption and/or an emission of a particle. A particle can also travel from one point to another, which is what we call propagation. Let us recall that in quantum theory an electron or a proton is detected as a particle, but is propagated as a wave. So different mathematical techniques are used to describe interaction and propagation.

A particle is a quantum of a corresponding field. So a photon is a quantum of an electromagnetic field, and an electron is a quantum of an electron field. A field is represented by the symbol $|n\rangle$ where the number n represents the number of quanta present in the field. If there are no quanta present, we represent the field by the symbol $|0\rangle$.

The symbol $|n\rangle$ is called a state function or state vector because it represents the state of the field, and the state of the field is described by the number of quanta present in the field.

The emission or the absorption of a quantum is represented by an operator — emission by the creation operator $a^\dagger$ and absorption by the annihilation operator a. Each one of these operators acts on some state. When we want to express the creation of a quantum in a field which previously had no quantum we use the expression $a^\dagger|0\rangle$. This describes the physical process of the creation or emission of a particle or quantum. The result of this creation is that there is now 1 quantum in the field. And so we can write the equation $a^\dagger|0\rangle = |1\rangle$. Likewise, we can express the annihilation or absorption of a photon by the equation $a|1\rangle = |0\rangle$.

We can check the field to ascertain the number of quanta present in the field. More precisely, we can ask questions such as: Is there no quantum in the field? Is there one quantum in the field? Are there two quanta in the field? And so on. Checking is represented by a state function written in reverse form. So the act of checking if a single quantum is present in the field is represented by the symbol $\langle 1|$.

The propagation of a particle is not described by a creation or an annihilation operator. So if the particle passes through an opening *but is not detected during this process* then this passage is described by a pair of operators: U which represents entry, and $U^\dagger$ which represents exit. If there is more than one opening, the propagation of the particle through the different openings is described by different operators such as $U(I)$ or $U(II)$, etc.

An action consisting of the creation of a particle and its subsequent absorption can be expressed as a symbolic narrative:

$$\langle 0|aa^\dagger|0\rangle$$

A more complex symbolic narrative could read as

$$\langle 0|aU^\dagger(I)U(I)|a^\dagger|0\rangle$$

This symbolic sentence has a numerical value which is a complex number having an absolute value between 0 and 1. If we know for certain that this sequence of events will happen, the sentence has the value 1. If we know for certain that this sequence of events could never happen, the sentence has the value 0. In general, there are many ways in which a particle — created at a point A — could be absorbed at a point B. Each of these ways gives rise to a different sentence. If there are two different possible paths that the particle could take between A and B we need to add up the values of the two sentences. This sum is called the amplitude for the event — particle created at A and detected or absorbed at B — to occur. The square of the absolute value of this amplitude (given by the amplitude multiplied by its complex conjugate) is the probability for this event to occur.

Chapter 11

Angular Momentum and Spin

11.1 Direction of the angular momentum vector

An object of mass m traveling along a circular path of radius r with speed v perpendicular to the radius of the circle has an angular momentum of magnitude mvr about an axis passing through the center and perpendicular to the plane of the circle.

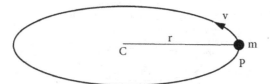

Particle P of mass m moving along a circle of radius r with center C

When a particle moves in a circle, its direction of motion changes continuously. But as long as the particle continues to move along a circle or ring the *plane* of its motion does not change. We could visualize a plane by thinking of a rectangular plate. Such a plate can be oriented in different ways in space. We artificially define the direction of a plane as the direction of a line drawn perpendicular to the plane. Since a circular motion takes place in a plane, the direction of the circular motion is defined as the direction of the axis passing through the center and perpendicular to the plane containing the circle.

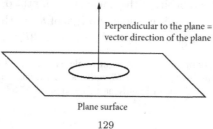

Perpendicular to the plane =
vector direction of the plane

Plane surface

129

There is an arbitrary rule for determining the positive direction. Suppose we look at the motion of the particle from a direction perpendicular to the plane of motion. If the motion of the particle appears clockwise, then the direction of the circulation is taken to be along the axis going *into* the plane of the motion. Another way of saying the same thing is to think of the circular motion as the rotation of a right-handed corkscrew. The linear direction along which the corkscrew advances when it is rotated is the direction of the circular motion considered as a single vector.

There are two quantities commonly associated with such a circular motion: angular velocity and angular momentum. The angular velocity is the rate at which the particle turns through an angular displacement. It is measured in radians per second.[1] It is usually represented by the Greek lower case omega ω. So $\omega = \theta/t$ where θ is the angle (in radians) that the object turns through in time t seconds.

The angular momentum has the magnitude of the product of linear momentum and radius, hence has the value mvr, and its direction is the same as the angular velocity.

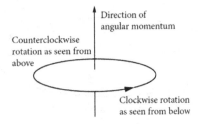

Direction of angular momentum

Counterclockwise rotation as seen from above

Clockwise rotation as seen from below

So when a quarterback spins a football in a clockwise direction he gives it an angular momentum in the forward direction. If a left-handed quarterback spins the ball in a counterclockwise direction he gives it an angular momentum in the reverse direction.

11.2 Quantization of angular momentum

We saw earlier (Ch. 7) that in Bohr's model of the atom the angular momentum of an electron orbiting the nucleus is quantized, i.e. the magnitude of the angular momentum is a whole multiple of $\hbar$. So the only possible values of the electron angular momentum are $\hbar, 2\hbar, 3\hbar, 4\hbar$, etc. We saw that

---

[1]A radian is a measure of an angle. A right angle equals 90^0. Expressed in radians, a right angle has the magnitude $\pi/2 = 1.57$ radians. A radian has no units. So 180^0 is commonly written as π radians or simply as π. 1 radian = 57.3^0.

this quantization of angular momentum is a consequence of de Broglie's law that every material particle has a wavelength associated with it and his hypothesis that the only permitted orbits are those that have a whole number of wavelengths in their circumference. $2\pi r = n\lambda$ where $\lambda = h/p$ and $p = mv$ is the linear momentum. Thus we get the equation: Angular momentum $mvr = n\hbar$.

When a particle is moving along a circle, its angular position is constantly changing. For a particle moving in a circle of radius r with center at the origin of a coordinate system, the angular position ϕ is defined as the angle made by the radius with the positive direction of the x axis:

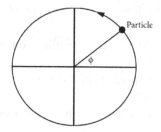

The uncertainty principle applies to angular momentum and angular position (just as it does for linear momentum and linear position). If we express the angular momentum by the letter L we may write

$$\Delta L \Delta \phi \gtrsim \frac{\hbar}{2}$$

The uncertainty principle had not been discovered when Bohr developed his model of the atom. This uncertainty principle requires a modification of Bohr's hypothesis. So the correct quantum theoretical understanding of the orbital angular momentum of an electron makes a distinction between measuring the total magnitude of the angular momentum and measuring the component of the angular momentum vector in some direction. The rules of quantum theory actually state that the *component* of the angular momentum measured along any direction should have values such as $-2\hbar, -\hbar, 0, \hbar, 2\hbar$, etc. The component of the angular momentum is an important quantity because in a real physical situation this component is a measure of the potential energy of the atom in a magnetic field. If the angular momentum is found parallel to the magnetic field, the component will be positive. If the angular momentum is in the opposite direction, the component will be negative. If the angular momentum is perpendicular to the magnetic field, the component is zero. But according to quantum theory, when the *magnitude of the angular momentum vector* is measured (i.e. not its component in any direction) we would get values such as $\sqrt{2}\hbar, \sqrt{6}\hbar, \sqrt{12}\hbar$, etc. One could think of the magnitude of the angular momentum as a measure of the rotational kinetic energy of the atom.

Every electron in an atom has a number ℓ called the *orbital angular momentum quantum number*, also called the *orbital quantum number*. The *magnitude* of the angular momentum is $\sqrt{\ell(\ell+1)}\hbar$ and the *component* of the angular momentum can have one of these values: $-\ell\hbar, (-\ell+1)\hbar, (-\ell+2)\hbar, ...0, \hbar, 2\hbar, ...\ell\hbar$.

Exercise 11.1.

1. Given the orbital quantum number of an electron to be $\ell = 1$ find

(a) the magnitude of the orbital angular momentum of the electron as an expression containing $\hbar$.

(b) all the possible components of the orbital angular momentum (as terms containing $\hbar$).

2. Given the orbital quantum number of an electron to be $\ell = 3$ find

(a) the magnitude of the orbital angular momentum of the electron as an expression containing $\hbar$.

(b) all the possible components of the orbital angular momentum (as terms containing $\hbar$).

A vector quantity is defined by its magnitude and its direction. While a vector has a definite direction, it also has a component in a direction different from its own, unless this different direction is perpendicular to the direction of the vector. In general, the component of a vector $\vec{A}$ in a direction at an angle θ to the vector is given by $A\cos\theta$. If $\theta = \pi/2 \ (= 90^0)$ then the cosine is zero and therefore the component is zero.

Quantum theory requires that the component of the angular momentum in any direction must be a whole multiple of $\hbar$: $\hbar, 2\hbar, 3\hbar$, etc. and these components could be positive or negative.

So we have an interesting situation. The measured value of the orbital angular momentum of an atomic electron is $\sqrt{\ell(\ell+1)}\hbar$ but the measured value of the component of the orbital angular momentum in any direction — say the direction of an external magnetic field — is $\ell\hbar$. This would be a contradiction in classical physics. In classical physics we could measure the components of the angular momentum along the three perpendicular directions, square them and add them up to obtain the square of the magnitude of the angular momentum:

$$L^2 = L_x^2 + L_y^2 + L_z^2 \qquad (11.1)$$

But such a procedure is impossible in quantum mechanics. The reason is that it is impossible to measure any two components (such as the x and the y components, the y and the z components, or the z and the x components) of the angular momentum *simultaneously*. There is an uncertainty principle (cf. Ch. 6) at work here similar to that between momentum and position.[2] But we can measure any one of these components (say L_z) as well as the magnitude of the angular momentum L simultaneously.

What we could do instead, is to take several *independent* measurements of the angular momentum in the x, y and z directions separately, find the average of the squares of L_x, L_y and L_z and add them up to get the average of the square of L. So Eq. (11.1) would become

$$\langle L^2 \rangle = \langle L_x^2 \rangle + \langle L_y^2 \rangle + \langle L_z^2 \rangle \qquad (11.2)$$

Let us apply this equation to the case $\ell = 1$. Each of the components of angular momentum can take on the possible values $\hbar, 0, -\hbar$. So

$$\langle L^2 \rangle = (\hbar^2 + 0 + \hbar^2)/3 + (\hbar^2 + 0 + \hbar^2)/3 + (\hbar^2 + 0 + \hbar^2)/3 = 2\hbar^2$$

So $\langle L^2 \rangle$ equals $2\hbar^2$ which is nothing but $\ell(\ell+1)\hbar^2$ for $\ell = 1$.

The total angular momentum of an electron in an atom is $\sqrt{\ell(\ell+1)}\hbar$ where $\ell = 0, 1, 2, 3$, etc. is called the *orbital angular momentum quantum number* or simply *orbital quantum number* of the electron. Suppose ℓ were to have the value 2. Then the total angular momentum is $\sqrt{6}\hbar = 2.45\hbar$ and the possible components of this angular momentum are $2\hbar, \hbar, 0, -\hbar$ and $-2\hbar$. The coefficient of $\hbar$ (i.e. the number 2, 1, 0, -1 or -2) in these numbers is called the *magnetic quantum number* because it becomes significant when the atom is placed in a magnetic field. The magnetic quantum number is expressed by the number m which is a positive, negative or zero integer.

The magnetic quantum number is manifested in the presence of a magnetic field. The angular momentum vector of the orbiting electron can only take on certain directions relative to the magnetic field. If we call the direction of the magnetic field the z axis, then the angular momentum of an electron with $\ell = 2$ can only take on these possible angles relative to the z axis: $\cos\theta = 2/2.45$ which makes $\theta = 35.3^0$, $\cos\theta = 1/2.45$ for which

---

[2] To be precise, the uncertainty principle forbids the simultaneous exact measurement of x and p_x, or y and p_y, or z and p_z. It does not forbid the simultaneous exact measurement of x and p_y, or y and p_z, or p_x and p_y, etc.

$\theta = 65.9^0$ and $\cos \theta = 0$ for which $\theta = 90^0$. The corresponding angles are the same for the negative values of m.

> **Exercise 11.2.** What are the possible angles between the angular momentum and the external magnetic field for an electron with $\ell = 1$?

11.3 Spin of an electron

An atomic electron has orbital angular momentum about the nucleus, and we can picture an atom somewhat like a solar system. Every electron also has angular momentum due to a feature called the spin of the electron. Here again, the analogy of the solar system comes to mind, but as with all analogies, this one too has limited value. The earth spins on its axis as it flies through space in a wide orbit round the sun. The axis of the earth's rotation is approximately perpendicular to the direction of its orbital motion. This macroscopic object could serve as a model for an atomic electron which has spin and linear motion as it orbits the nucleus. However, when we consider a microscopic particle such as an electron, it is not possible to think of its spin as a rotation about its axis because an electron is really not a hard spinning sphere of charge. Indeed, it is not possible to picture the electron at all, because any attempt at picturing involves a mental construction of its shape and trajectory of motion, neither of which is real when we are dealing with extremely small objects. Nevertheless, there is some merit to thinking of the spin of an electron as a rotation about its axis.

The law of conservation of angular momentum states that the total angular momentum of a system of objects when measured about any axis remains constant unless there is an external rotational force (more properly, a *torque*) acting on the system. It turns out that for an atomic electron only the sum of its orbital angular momentum and its spin angular momentum is conserved, not the spin or the orbital angular momentum separately. So it is possible for the electron spin to undergo a spontaneous flip (angular velocity vector changing direction from one direction to the opposite), but if that should occur there would also be a change in the orbital angular momentum, maintaining the total angular momentum constant. Thus there is some value in picturing spin as a rotation. A second argument in support of this can be produced from electromagnetism. In classical physics a spinning charge behaves as a magnet with a north and a south pole. In an iron atom, the spin of one of its outer electrons causes the atom to behave as a tiny magnet. Since the magnetism of the atom comes from the spin of a charge, it makes sense to think of spin as a rotation. Thus it is logical to picture the electron as a spinning charge, all the while keeping in mind that in quantum theory a spinning particle is no more than a convenient symbol.

The clearest evidence that the spin of an electron should not be thought of as a spinning top is the behavior of an electron in a magnetic field. If the electron were a rotating charged sphere, the magnetic field would cause the electron to precess about the direction of the magnetic field, i.e. the axis of rotation of the electron would rotate about the direction of the magnetic field:

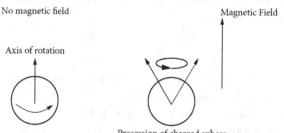

Precession of charged sphere
Axis of rotation of sphere rotates about the magnetic field

Moreover, the angle of precession could have any possible value.

But the electron spin does not behave that way. The electron spin axis can have only two possible directions relative to the magnetic field: parallel or antiparallel (opposite):

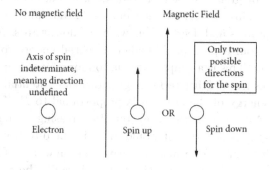

11.3.1   *Stern–Gerlach experiment*

A negative electron with spin aligned parallel to the magnetic field has a higher energy than one with spin antiparallel to the field. Based on this fact, an experiment can be performed that demonstrates that an electron spin can only be parallel or antiparallel to a magnetic field. This is done by sending a stream of electrons through a magnetic field. If the direction of the spins is entirely random before going through the magnetic field, there is a 50 per cent probability that the spin gets aligned parallel to the field and a 50 per cent probability that the spin gets aligned antiparallel to the field. Let a stream of electrons be sent along the x axis, and let us apply a magnetic field in the z direction. If the strength of the magnetic field has magnitude B, then the potential energies of the two possible alignments of the electron spin are given by

$$\frac{e\hbar B}{2m} \quad \text{and} \quad -\frac{e\hbar B}{2m}$$

where e and m are the electron charge and mass respectively, and $\hbar = \frac{h}{2\pi}$. The positive potential energy corresponds to the negative electron having spin in the same direction as the magnetic field.

It would be impossible to distinguish between electrons in the spin up or spin down state without physically separating the two states. Since the two states have different energies, we could use their difference of energies to find a way of splitting the beam so that electrons in one state go one way, and electrons in the other state go another way.

One way of splitting the beam is to introduce a non-uniformity in the magnetic field, by giving the field a gradient or a continuous increase along the z direction. This means that the value of B is not uniform but increases as we move upwards and decreases as we move downwards. Now, particles tend to move from a position of higher potential energy to one of lower potential energy. This is a simple explanation for why an apple falls from a tree — it moves from higher to lower gravitational potential energy. Since the potential energy of the electron is proportional to B, if the energy is positive, it would move in the direction of decreasing B, i.e. downwards. And if the energy is negative, it would move in the direction of increasing B (to make the energy even more negative), i.e. upwards. Thus, electrons in the two states are now separated into two diverging beams as they pass through this non-uniform magnetic field.

Such an experiment was carried out by Stern and Gerlach. They showed that the electron beam was split into two streams as it passed through the apparatus. Classical physics predicts that the electron beam would fan out in both directions with electrons being deflected through a continuous range of angles, because in classical physics the electron spin can have any orientation relative to a magnetic field. But the experiment showed only two directions, thereby demonstrating that the spin had only two orientations relative to the magnetic field. Thus the Stern–Gerlach experiment provides experimental proof for the quantum theory of electron spin.

11.4 Pauli exclusion principle

Precise calculations verified by experiment show that particles have intrinsic angular momentum characterized by a number conventionally denoted by the letter s which can be $0, \frac{1}{2}, 1, \frac{3}{2}, 2, \frac{5}{2}, 3$, etc. and the intrinsic angular momentum — called *spin angular momentum* — is given by the expression

$$\sqrt{s(s+1)}\hbar$$

The number s is called the spin of the particle. An electron has spin $\frac{1}{2}$, and the W and the Z bosons have spin 1. The highly unstable Higgs boson has zero spin.

The intrinsic spin of an electron has some effects that are similar to those produced by a rotating charged body. One of these effects is magnetism, since a rotating charged body can be considered a charge moving in a circle, or a current moving in a circular loop, which is basically a magnet (Ch. 4).

As with the case of orbital angular momentum, the spin angular momentum of an electron can have only certain orientations relative to a magnetic field. These orientations are either parallel to the field or antiparallel to the field, represented by the different numbers $1/2$ and $-1/2$, and which has been confirmed by the Stern–Gerlach experiment. A particle in a stable state such as an electron in an atom can have only discrete values of total angular momentum, total energy, etc. A particular set of these values collectively expresses the *quantum state* of the particle. The quantum state is also expressed by quantum numbers. For example, the ground state of the atom is expressed by the energy quantum number $n = 1$. The first excited state has energy quantum number $n = 2$. In addition to energy and angular momentum, an atomic electron also has some other quantum numbers depending on how the orbital angular momentum and how the spin angular momentum are oriented in the presence of an external magnetic field. So the quantum state of an atomic electron is described by a set of quantum numbers that represent the different physical quantities that can be measured for the electron. If two different electrons in the same atom could have all the same quantum numbers, then they would be in the same state. But this cannot happen by the *Pauli exclusion principle.*

Particles having spins which are odd multiples of $\frac{1}{2}$ obey the Pauli exclusion principle, which states that two such identical particles cannot be in the same state. This means they cannot have all the same quantum numbers in any system. Suppose we have two different hydrogen atoms at a large distance from each other. Suppose each of them has an electron in the ground state. It is entirely possible that both these electrons have the same set of quantum numbers. They appear to be in the same state, but they are in different systems. The situation changes if the two hydrogen atoms bond together to form a hydrogen molecule. This is a more complex system with a more complex set of quantum numbers. This system has two electrons, and these two electrons cannot be in the same state.

The Pauli exclusion principle helps explain the chemical properties of elements. Because no two electrons have all the same quantum numbers, for an atom with several electrons it is necessary to assign unique quantum numbers to each electron. Hence it is not possible to put all the electrons in the lowest energy state corresponding to $n = 1$. Some electrons will have $n = 2, 3$ etc. depending on the size of the atom.[3] The chemical properties of an element generally depend on the quantum numbers of the electrons having the highest value of n, though the actual details are somewhat complex. Carbon has six electrons, two electrons having $n = 1$ and four electrons with $n = 2$. Nitrogen has seven electrons, two electrons having $n = 1$ and five electrons with $n = 2$. Nitrogen and carbon have vastly different physical and chemical properties, largely because of the difference in the number of electrons having $n = 2$ in the atoms. An element is defined by the number of electrons in the atom which is equal to the number of protons in the nucleus. Every nucleus (except the commonest form of hydrogen) contains neutrons in addition to the protons. The neutrons cause the nucleus to be stable. But since the neutrons have no charge, they do not affect the number of electrons in the atom which always equals the number of protons. So if two different atoms have the same number of protons but have different numbers of neutrons they would have the same chemical properties. Different forms of the same element having different numbers of neutrons are called *isotopes* of the same element.

---

[3]Helium has two electrons, which are normally both in the lowest energy state with $n = 1$. But they have opposite spins, and so are not in the same state. Lithium has three electrons, and the third electron has $n = 2$.

11.4.1 *Quantum statistics*

Earlier we saw that the study of the properties of a body — solid, liquid or gas — in terms of the mechanics of the individual molecules that constitute the body is called statistical mechanics (Ch. 3). Prior to the advent of quantum theory, the physical behavior of substances in terms of their molecular constitution was studied largely by Maxwell and Boltzmann and so this classical statistical mechanics is called Maxwell–Boltzmann statistics.

The distinctive feature of Maxwell–Boltzmann statistics that sets it apart from quantum statistical mechanics is that in Maxwell–Boltzmann statistics the individual molecules are treated as distinguishable particles. Each molecule has its own distinct individuality. The situation is vastly different with quantum statistics.

Such particles having spin $\frac{2n+1}{2}$ where $n = 0, 1, 2, 3, \ldots$ are called *fermions*, because their collective or statistical behavior was studied by Fermi and Dirac, and is called Fermi–Dirac statistics. Because electrons have spin 1/2, they are fermions, and obey the exclusion principle. We cannot place two electrons in the same quantum state. This is one of the reasons why we do not fall through the floor. The molecules on the soles of our shoes are in close contact with the molecules on the surface of the floor. This means that the electrons of our shoes are pushed up close against the electrons of the floor. So the weight of our bodies due to gravity causes some electrons to come very close to one another, and if they came too close, they would be occupying the same quantum state. Since this must be avoided at all costs, our feet experience a strong force of repulsion from the floor, which keeps us firmly rooted on the floor without going through it. The exclusion principle is also responsible for resistance in an electrical circuit. Electrons are pushed apart from each other not only because like charges repel but more especially because they resist being forced to occupy identical quantum states. However, certain conductors at very low temperature lose their resistance altogether. This is because electrons form pairs with opposite spins so that each pair behaves like a particle with zero spin. Such electron pairs are called *Cooper pairs* and this phenomenon is called *superconductivity*. Particles with spins that are whole numbers are not subject to any exclusion principle. One can place any number of such identical particles in the same state. Such particles having spin n where $n = 0, 1, 2, 3 \ldots$ are called *bosons* because their statistical behavior was studied by Bose and Einstein, and is called Bose–Einstein statistics. Photons are bosons because they have spin 1. An unlimited number of photons with the same frequency can be placed on top of each other, and this is what takes place in a laser beam. Another boson is a Helium 4 nucleus. The two protons and the two neutrons of a helium nucleus are oriented such that their net spin is zero, and so such a nucleus behaves like a spin 0 particle and so is a boson. Also the two electrons in a helium atom have opposite spins. At very low temperatures helium loses its viscosity altogether and becomes what is called a *superfluid*. Ordinary liquids have viscosity because the molecules cannot pass through each other and so a flowing liquid feels an internal resistance due to the friction between its molecules. If the molecules could all occupy the same quantum state then they could be placed on top of each other or could move through each other. This is what happens in superfluidity. Another isotope of helium has two protons and one neutron (He^3) and since the spins cannot balance each other, this nucleus is a fermion. But as in the case of electrons in a conductor, He^3 nuclei can also form pairs with opposite spins and thereby behave as bosons. So superfluidity has also been observed in liquid Helium 3.

11.5 Summary

The angular momentum vector of a rotating object is directed along the axis of rotation. If the rotation is clockwise when viewed along a direction perpendicular to the plane of rotation the angular momentum vector is directed away from the observer.

In quantum theory we cannot picture the trajectory of a moving particle. So it is impossible to provide an accurate picture of an electron orbiting round a nucleus in an atom. But it is possible to measure the angular momentum of this orbital motion along any particular direction. This angular momentum measured along an arbitrary direction will always be an integer multiple of $\hbar$: positive, zero or negative. The orbital motion of an electron in an atom is characterized by a whole number ℓ called the orbital quantum number. When the angular momentum is measured along some direction, the angular momentum we obtain is always an integer m times $\hbar$. The number m is called the magnetic quantum number of the electron's orbital motion. m can take on a total of $2\ell + 1$ values, viz. $\ell, \ell - 1, \ell - 2, \ell - 3, ... 3 - \ell, 2 - \ell, 1 - \ell, -\ell$. One can also measure the square of the angular momentum of the electron, which would come out to be $\ell(\ell + 1)\hbar^2$. So the magnitude of the orbital angular momentum would be $\sqrt{\ell(\ell + 1)}\hbar$.

In addition to the orbital angular momentum, every electron has a spin angular momentum characterized by the spin quantum number $\frac{1}{2}$. If this angular momentum were measured in any direction, one would obtain either $\frac{1}{2}\hbar$ or $-\frac{1}{2}\hbar$. The magnitude of the spin angular momentum is $\frac{\sqrt{3}}{2}\hbar$.

Every particle has a spin quantum number. This number could be a half integer such as $\frac{1}{2}$ or $\frac{3}{2}$, etc. Or it could be a whole number such as 0, 1, 2, etc. Particles with half integer spins are called fermions because when they are placed with other identical particles they follow a statistical rule of behavior called Fermi–Dirac statistics. Particles with whole number spins are called bosons because when they are placed with other identical particles they follow Bose–Einstein statistics. Fermions follow the Pauli exclusion principle whereby two identical fermions cannot be in exactly the same place, which means they cannot have the same set of quantum numbers. Bosons do not obey any exclusion principle.

Chapter 12

Quantum Theory and Relativity

12.1 Dirac theory

12.1.1 *Negative energy states*

According to Einstein's Special Theory of Relativity (Ch. 8), the mass of a moving object m is related to its rest mass m_0 by the equation

$$m = \frac{m_0}{\sqrt{1 - \frac{v^2}{c^2}}}$$

The energy of this object becomes

$$E = mc^2 = \frac{m_0 c^2}{\sqrt{1 - \frac{v^2}{c^2}}}$$

Using the definition of momentum $p = mv$ we obtain with some algebra

$$E^2 = p^2 c^2 + m_0^2 c^4 \qquad (12.1)$$

This equation relates the total energy E of a particle to its momentum p and its rest mass m_0.

Exercise 12.1. Derive Eq. (12.1).

What is interesting about this equation is that the energy and momentum appear only in the second power. This means that for any particular value of the momentum the energy can have two values, one positive, and the other negative:

$$E = \pm\sqrt{p^2 c^2 + m_0^2 c^4} \qquad (12.2)$$

Negative energies are common in physical situations. We saw in Ch. 7 that for an atomic electron the potential energy is negative and greater in

143

magnitude than the positive kinetic energy, and so the total energy comes out negative. Thus, negative energies are not unphysical. Now, for a free electron that is not attached to any atom the potential energy is zero and so the total kinetic energy has always been assumed to be positive. But now Dirac derived the surprising result that negative energies are possible even for a free electron.

But this raised a problem. If negative energy states are real, how come every electron does not drop into a negative energy state in a manner analogous to an atomic electron dropping from a higher to a lower energy state? If this could happen, there should be no positive energy electron available in the universe as they would all drop from higher (positive) to lower (negative) energy states. One way out of this problem was to assume that the negative energy states were all filled. This would be analogous to an atom in which the lower energy states are occupied by electrons, and by Pauli's exclusion principle (Ch. 11) the higher energy electrons could not drop to one of these lower energy states since no two electrons can occupy the same quantum state. So Dirac's hypothesis was that the universe is actually an enormous sea of negative energy electrons of all possible (negative) energies, and so there is no possibility of a positive energy electron dropping to a negative energy state, simply because there is no room available.

But in an atom it would be possible for a higher energy electron to drop to a lower energy state if somehow one of the lower energy electrons got knocked out of the atom, thereby creating a "hole". An atom with such a "hole" would have a net positive charge, because the positive charge of the nucleus would not be canceled by the negative charges of all the remaining electrons. A positively charged ion is just such an atom which has lost one or more of its electrons.

So a hole in the sea of negative electrons would behave as a positively charged particle. Dirac proposed that such a particle must exist. And so the idea of a *positron* was born. (The name came later.) The positron was eventually discovered. Positrons are given off by the nuclei of certain atoms. We saw earlier that positrons collide with electrons and the two oppositely charged particles annihilate to produce a pair of photons. Because these positrons die so quickly after they are generated by the nucleus, they are difficult to detect.

Today positrons are employed in the medical profession in *Positron Emission Tomography* (pet) scanning of organs within the human body. Compounds are made of molecules, and molecules are made of atoms. A compound containing an atom that is capable of generating positrons is injected into the body. These compounds attach themselves to certain cells, and from there they emit positrons. These positrons quickly annihilate with surrounding electrons and two photons are emitted at each annihilation. These photon pairs generate a picture of the organ or part of the body that is being examined. Such a picture is called a *pet scan*.

12.1.2 *Antiparticles*

The positron is called the antiparticle of the electron, and vice versa. Indeed, every particle in nature has a corresponding antiparticle. The antiproton has the same mass as a proton but a negative charge. This particle has also been detected, but like the positron, it does not survive for very long after creation because it is rapidly annihilated upon collision with a proton.

Neutral particles too have their antiparticles. A neutron is neutral. A neutron has *baryon number* +1 and an antineutron has baryon number −1. The two particles differ in their fundamental structure. Many particles are composed of constituents called *quarks*. Neutrons and antineutrons differ in their quark structure.

A positron orbiting an antiproton would constitute an atom of antihydrogen. Such an atom has been created in the laboratory, but it is difficult to keep it in existence for any length of time, because it would be annihilated when it collides with regular matter. An antiproton would be annihilated upon collision with a proton, and the positron likewise with an electron. There is no reason in principle why there cannot be antihelium, antioxygen, anticarbon, etc. Such elements constitute what is called *antimatter*. Antimatter has positive mass and obeys all the laws of physics that we know. One of the puzzling features of the observable universe is that it is apparently asymmetric with respect to matter and antimatter. Our galaxy seems to be made exclusively of atoms with protons, neutrons and electrons. In principle there could be galaxies made of antiprotons, antineutrons and positrons. But we have no evidence that such galaxies actually exist.

12.1.3 *Zitterbewegung*

We saw that a positron can be thought of as a hole created when a negative energy electron is raised to the positive energy state. Thus a positron is NOT a negative energy electron. Dirac thought it was impossible for an electron to fall into a negative energy state because all the negative energy states were full, and that the only way a negative energy state could be emptied was by the creation of a hole, which he interpreted as a positron, and that is a rare phenomenon. So Dirac assumed that a positive energy electron would rarely drop to a negative energy state. But Dirac was not entirely correct in this matter.

Schrödinger used Dirac's equation to show that a free electron constantly oscillates between positive and negative energy states. He called this oscillation *zitterbewegung*, which denotes a trembling or jittery motion. The significance of this term is that in zitterbewegung the electron does not only oscillate between energy states, it also undergoes spatial motion. A stationary electron is therefore not quite stationary. It is displaced from a point A to another point B very close to A and again back to A. The surprising thing is that this displacement seems to take place at the speed of light. But we know that an electron cannot really travel at the speed of light. At any rate, since the electron — like every microscopic particle — does not have a trajectory, we should not picture the electron flying between A and B at the speed of light.

The frequency of this jittery motion is so very high that it is impossible to observe it in the laboratory. However, analogous phenomena have been discovered for particles much bigger than electrons, and these bigger particles have slower zitterbewegung frequency. So zitterbewegung as a quantum phenomenon has been confirmed.

12.2   Entangled states

We saw in Ch. 7 that every fundamental particle has a wavelength inversely proportional to its momentum. We also saw that a collection of particles such as an atom or molecule also has a wavelength related to its momentum — according to the same de Broglie formula. In the latter case the constituent particles of the atom or molecule are entangled, so that the collection of particles behaves like a single object.

Entanglement can also be performed between two electrons. Suppose we were to entangle two electrons having opposite spin directions. Then their net angular momentum would be zero. And this would be true no matter what be the axis about which the angular momentum is measured. We have seen that magnetism in iron is mainly due to the spin of the atomic electrons. Thus a spinning electron is a tiny magnet with a magnetic moment. (Magnetic moment is a measure of the strength of a magnet.) If the two electrons that we have entangled have opposite spins, their combined magnetic moment is zero.

Now, suppose this pair of electrons were to pass through a magnetic field, say along the z axis of our coordinate system, and we try to measure the magnetic moments of the electrons separately. This is possible in principle. Since an electron would always be found with its spin parallel or antiparallel to the magnetic field, we would get the following interesting result. Suppose the first electron we measured was found with spin parallel to the field, i.e. along the positive z direction. Then the next electron would automatically be found with spin antiparallel to the field, i.e. along the negative z direction, so that the net spin of the two-electron system is zero.

Next, suppose we pass the pair of electrons through a different magnetic field along the x axis. One electron would be found with spin along the positive x direction and the other electron with spin along the negative x direction.

Here comes the bizarre prediction of quantum theory.

Suppose the two entangled electrons are sent off in opposite directions. They would travel away from each other, but *their entanglement would be unbroken!* What does this mean? Let us say these electrons are millions of kilometers from each other, one on the planet Mercury and the other on the planet Neptune. Let us say prior to sending these electrons away from each other, two astronauts — with synchronized clocks — had set out (several years earlier) to these two planets. Astronaut Alice is on Mercury and astronaut Bob is on Neptune. Suppose the two electrons reach Alice and Bob at the same time as measured by their clocks. Each astronaut measures the spin of the electron along a direction that is parallel to the axis of rotation of the sun, and therefore these directions are the same for both Alice and Bob. The results of the two measurements would be ONE of the following:

(a) Alice finds her electron with spin up, and Bob finds his electron with spin down.

(b) Alice finds her electron with spin down, and Bob finds his electron with spin up.

Since these are the only two possibilities, we can state the following with absolute certainty:

If Alice finds her electron has spin up, then Alice knows for certain that Bob will find (or has found) his electron with spin down.

So by measuring the spin on her electron, Alice obtains instant information about another event that is several light hours away.

12.3   Apparent conflict with relativity

12.3.1   *Action at a distance*

This is a natural consequence of quantum mechanics. But it is amazing from the point of view of Special Relativity. It seems to imply that Alice's act of measuring her electron has influenced Bob's electron instantaneously, or the other way round. It would appear that a message has been sent from Alice's electron to Bob's electron or vice versa at infinite speed, or at any rate faster than the speed of light.

Naturally, Einstein was not pleased with this result from quantum theory. He called this phenomenon "spooky action at a distance" and refused to believe it. He devised some really clever arguments to refute this aspect of quantum theory. It took the best minds in the world to show where he was wrong.

We had earlier discussed the phenomenon of action at a distance mediated by gravitational and electromagnetic fields (Ch. 4). We now take up this issue again and discuss it in the context of entanglement.

An instantaneous interaction at a distance is called a non-local interaction. We saw earlier that non-local interactions were assumed to be natural until as late as the 19th century. The Greek philosopher Archimedes is credited with this claim: "Give me a lever and a place to stand and I will move the earth." The lever is a simple machine where an effort applied at one end is able to raise a load at the other end. If the effort arm is much longer than the load arm, then a small effort can lift a large load. Archimedes believed that there is no limit to the load that could be raised provided the ratio of the effort arm to the load arm (called velocity ratio) was sufficiently large. Thus the mechanical advantage (load divided by effort) could be made arbitrarily high.

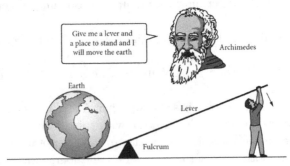

Archimedes was of course assuming that the extremely long lever was perfectly rigid. A perfectly rigid lever would not bend under any force. Thus an effort applied at one end would be communicated instantly to the other end, lifting the load without any time lapse between the application of the effort and the raising of the load. Thus perfect rigidity implied the possibility of communicating information and energy at speeds exceeding that of light. But perfect rigidity is impossible. All solids are made of atoms, and the rigidity of solids arises from the very strong electronic bonds between neighboring atoms. These bonds are not infinite in strength, which allows even "iron bars to bend and break." And the bending force takes time to travel from one end of the rod to the other. So when an effort is applied to one end of a lever, the turning force is communicated through the lever from one end to the other via the atomic bonds of the material. And this bending force travels as a wave through the body of the lever at a finite speed that can be calculated from the physical properties of the material of the lever. So if the lever were made of iron, the wave would travel at about 6 km/s, which falls way short of the speed of light.

But the atomic theory of matter — even though it had been suggested by some philosophers in antiquity — was not universally accepted until it was experimentally proved through observations such as the Brownian movement and its interpretation. Solid state physics uses quantum mechanics to calculate the speed of propagation of disturbances through solids, and this development did not take place before the 20th century. Newton's theory of gravitation assumed that gravitational forces were communicated instantaneously from one object to another. Whatever misgivings Newton himself may have had about this idea, it appears that 18th century physics had no real alternative to action at a distance.

12.3.2 *Action mediated by a field*

But the turning point came in the 19th century when the classical notion of interactions at a distance was replaced by the concept of action mediated by a field in Maxwell's electromagnetic field theory. In this theory — called Classical Electromagnetism — all interactions are mediated by the electromagnetic field, and no influence can travel faster than the speed with which electromagnetic waves propagate through space, which is also the speed of light 3×10^8 m/s, as we saw in Ch. 4. So in contrast to Newtonian physics electromagnetic field theory is a local theory. A charge A interacts with a local field, which then communicates the influence to a distant charge B.

Thus the action between A and B is mediated by a field. At every point in this process the interactions are localized. Einstein developed his theory of Relativity building on the results of Maxwell's theory of electromagnetism. Einstein extended the idea to gravity as well. Hence the gravitational attraction between the earth and the sun is mediated by the gravitational field due to the earth and the sun. If the sun were to blow up suddenly, the effect would not be felt on the earth till the influence traveled from the sun to the earth, and these gravitational waves also travel at the speed of light in Einstein's theory. And so according to Relativity all physical interactions had to be local. And so the latter half of the 19th century and the early decades of the 20th century saw the paradigm of action at a distance replaced by action mediated by fields. In the quantum theory of electromagnetism all forces between charged bodies are mediated by fields — and more precisely by quantized fields which consist of photons traveling at the finite speed of 3×10^8 m/s. Likewise forces between the nuclear particles such as protons and neutrons are communicated by the quanta of the strong field — called mesons — which also travel at finite speeds.

Thus, the surprising non-locality of quantum entanglement came as a jolt to many physicists, a group that included Einstein as its most illustrious and most vocal member. But Einstein's misgivings were probably unfounded, because quantum non-locality does not really violate the Special Theory of Relativity.

12.3.3   *Communication of information*

The reason is that no information can be sent faster than light through quantum entanglement. Specifically, in this situation Alice cannot send to Bob any information about her measurement. All that Bob knows is that when he measured his electron he found its spin in the downward z direction. Now, any spin has a chance of being up or down along any axis. So after his experiment Bob can say almost nothing about Alice's measurement. He cannot even infer that Alice has made a measurement at all. The only thing he can say for certain is that Alice could NOT have measured her electron and found its spin to be exactly along the downward z direction. Equivalently, he could be certain that IF Alice had measured her electron spin along the z axis it would have come out to be in the upward direction.

Now, if Alice and Bob told each other that they would perform their measurements along the y axis, and if each believed the other, then the moment Alice measured her electron along the y axis she would know that Bob's measurement came out in the opposite direction as hers. But this is not a faster-than-light communication between Bob and Alice. It is an inference based on something that has very little to do with the laws of physics — Alice's trust in Bob's integrity, that Bob indeed measured his electron along the y axis as he said he would.

Suppose Alice were a bit more venturesome. She tells Bob she is going to measure along the z axis but instead she measures it along the x axis. Is there any way that Bob would know this by carrying out a measurement on his electron?

Suppose Alice measured the spin of her electron in the x direction and that it came out in the negative x direction. Remember that an electron spin will always align itself either parallel or antiparallel to the magnetic field that is used to measure the electron spin. So if Bob measures his electron spin in the z direction, it would come out with a 50–50 chance of being up or down in the z direction. So Bob would infer that Alice's electron had spin opposite to whatever spin direction he obtained in his measurement. There is no way that Bob could know that Alice had measured her electron in the x direction. But of course Bob's electron would definitely have spin in the positive x direction if by some whim he decided to do his measurement also along the x axis. But since he has no means of knowing that this result was inevitable he could still assume that Alice had measured her electron in the z direction.

To conclude, by entangling two particles and separating them, the result of a measurement on one particle is directly linked to the result of a measurement on the other, but this does not enable one observer to send any information to the other. And so the causality of relativity is not violated.

The apparent violation of relativistic locality is sometimes stated in terms such as these: "A measurement made on one of a pair of entangled particles *immediately* affects the other particle." This wording might convey the (unintended) impression that somehow some *information* traveled at infinite speed from one particle to the other. But, as we have seen, that is not the case. There is no instantaneous communication of information.

But a real weakness with the formulation quoted in the previous paragraph is that it implies *a causal relationship between the two measurements*. It is as if first one measurement is done on one particle, which then immediately places the second particle in a constraint, so that when the second measurement is done the outcome has been determined by the outcome of the first measurement. It is not hard to show that this causal relationship becomes meaningless when both observers are in relative motion.

12.4   Time ordering of measurements is relative

Let us return to the case of the two trains passing each other with very high relative velocity that we discussed in Ch. 8. Train I is moving in the opposite direction as Train II and both are running along adjacent parallel tracks.

Suppose Alice and Bob are at the two ends of Train I. Alice is at the engine, and Bob is at the caboose. Charlie is at the center, and he prepares a pair of electrons entangled in their spins having total spin zero. And Charlie has measured the total spin of the pair and found this sum to be indeed zero. He has not made any measurement on the individual spins of either electron. He sends one electron to Alice at one end of the train and the other to Bob at the other end. Both Alice and Bob have received their electrons. They do not carry out their measurements yet. They wait for Train II.

Alice has determined that she will measure her electron the moment she passes the caboose of Train II. She has a firing system directed onto Train II that will fire an X if her electron is found with spin up, and an O if her electron is found with spin down. So the caboose of Train II will have a mark of one or the other shape depending on the outcome of Alice's measurement.

Bob has determined that he will measure his electron the moment he passes the engine of Train II. He too has a gun that will fire an X onto the engine of Train II if he finds his electron to have spin up, and an O if his electron is found to have spin down.

Charlie, who is in the center of Train I, records the observations of Alice and Bob. He observes that Alice and Bob get opposite results for their measurement of the spins of the electrons in the z direction.

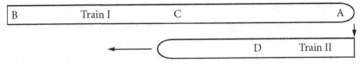

In Charlie's reference frame Alice does the measurement first,
because she passes the caboose of the other train first.

Within Train II right at the center sits an observer Delicia. Delicia records the two measurements made on Train I, and observes the engine and the caboose of her own train (Train II). She notes down the shapes of the marks at the two ends of her train, and sees that they are different.

Let us assume that both trains have the same length when they are stationary. But when they are in relative motion with relative speed v, each train will see the other train as shortened by the relativistic contraction corresponding to speed v. So if the rest length of each train is L_0, the length of each train as measured by a passenger in the other will be

$$L = L_0 \sqrt{1 - \frac{v^2}{c^2}}$$

This means Charlie will find Train II shorter than his own train, and Delicia will find Train I shorter than hers. This is one of the consequences of the Special Theory of Relativity that we studied in Chs. 8 and 9.

How will this impact the observations made by Charlie and Delicia?

Since Charlie will see Train II shorter than his own Train I, he will observe his engine passing the caboose of Train II before his caboose passes the engine of Train II. So he will observe Alice's measurement earlier than Bob's. So according to Charlie, Alice performs her experiment first, and finds her electron to have a particular direction of spin. Let us say she finds her electron with spin up in the z direction. And she does this as soon as she passes the caboose of Train II, and marks the caboose of Train II with an X. Since Train II is shorter than Train I in this frame of reference, it will take some time for the engine of Train II to reach the caboose of Train I. When it does so, Bob will perform his experiment, and we know that he will find his electron with spin down, and so he will fire an O onto the engine of Train II. So Charlie will conclude that Alice's observation immediately placed Bob's electron in a spin down position. Alice's measurement was the cause, and the *result* of Bob's measurement the final effect.

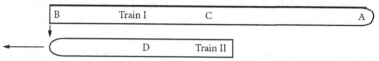

In Charlie's reference frame Bob does the measurement after Alice has done hers, because Bob passes the engine of the other train after Alice has passed the caboose of that train.

But a different causality is observed by Delicia. From her point of view Train I is shorter than Train II. She will therefore observe her engine crossing the caboose of Train I first. So she will observe Bob's measurement first.

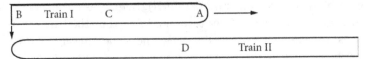

In Delicia's reference frame Bob makes the first measurement because he passes the engine of the other train before Alice reaches the caboose of that train.

She will therefore see the O mark landing on her engine first, and a little later, when the engine of Train I reaches her caboose she will see the X mark from Alice's gun. So for Delicia Bob's measurement is the cause, and the result of Alice's measurement the effect.

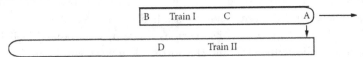

In Delicia's reference frame Alice does her measurement after Bob has completed his, because Alice reaches the caboose of the other train after Bob has passed the engine of that train.

Thus it is impossible to determine in an absolute way whether it was Alice's measurement that affected Bob's result or the other way round. One thing is certain. It is misleading to say that when Alice measured her electron her action immediately affected Bob's electron. For this would make Alice the agent and Bob the object in an absolute sense. And we have seen that the roles of agent and object can be reversed in different frames of reference. So it does not really matter who measures first. All that matters is that the results of the two measurements — whether they are seen as simultaneous or at different times — are not independent for entangled particles. It is wrong to say that one outcome *has an effect on the other*. Thus relativistic causality is not violated. It is indeed impossible to send any sort of message faster than the speed of light, even using entangled particles.

Nevertheless, even though relativity is not violated, the fact that quantum theory is a non-local theory could still leave many people with an uncomfortable feeling, as it did Einstein and many others who shared his worldview of physics.

There is a way of reconciling both the results of quantum theory and the need for preserving the "local" nature of physics. But this involves time travel.

12.5 Feynman graph of entanglement and measurement

We shall now draw a Feynman diagram (cf. Ch. 9) of the process we have been discussing so far. Charlie prepares an entangled pair of electrons with net spin angular momentum zero. He sends one electron to Alice and the other to Bob. Alice measures her electron and Bob measures his electron. If Alice found her electron to have spin up along any axis, then Bob would find his electron to have spin down along the same axis.

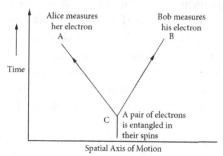

We have seen that time is reversible for a process involving a single particle. So an electron moving forwards in time can be thought of as a positron moving backwards in time and vice versa. Quantum theory does not distinguish between these two processes.

So we could interpret the Feynman graph shown above either as (a) one electron traveling from C to A and another traveling from C to B, with both electrons moving forwards in time, or as (b) a positron traveling backwards in time from A to C, and an electron traveling forwards in time from C to B.

Both interpretations are fully equivalent, and are physically identical. But the second interpretation might help to salvage the notion of locality.

Let us say Alice performs a measurement on her electron and finds it to have spin up in the z direction. This information is carried backwards in time by the electron — now acting as a positron — till it meets the second electron at C. Now Charlie ensures that the net spin angular momentum of the pair of particles is zero. So that immediately constrains the second electron to have a spin exactly opposite to that measured by Alice. Thus the spin as measured by Bob at B was already determined at C. Bob's measurement is therefore the direct consequence of Alice's measurement according to this picture.

Of course, this is a reversible model. One could have made Bob the primary experimenter. His measurement on the electron spin is sent back in time to C where it constrains the other electron so that when Alice measures her electron she can get only one result.

According to this picture the theory is fully local. All influences travel strictly along the world lines. There is no "spooky action at a distance."

12.6   Summary

The basic laws of quantum theory are not changed when the Theory of Relativity is taken into account. But the application of Relativity to quantum theory yields some surprising results. In particular, when Relativity is applied to the quantum theory of an electron it opens up the possibility that the electron could be in a negative energy state. This led to the hypothesis that there exists a particle identical to the electron except that its electric charge is equal and opposite to that of the electron. This particle is called the positron.

An electron oscillates constantly between positive and negative energy states, and simultaneously undergoes a jittery motion in space. This is called zitterbewegung. Since the frequency of zitterbewegung is very high, it cannot be detected in the laboratory. But zitterbewegung has been observed in particles much heavier than an electron, which have smaller frequencies.

The positron is the antiparticle of the electron. Every particle has its own antiparticle. An atom composed of antiprotons, antineutrons and positrons is called an atom of antimatter. It is theoretically possible for entire galaxies to be made of antimatter but we have not detected any so far.

Certain properties — such as the spin angular momentum — of two or more particles can be linked in such a way that the property remains a constant even when the individual particles move far apart. In the language of quantum mechanics we say that entangled particles are described by a single quantum state. And this state does not change even when the particles are not in physical proximity with each other.

So if the spins of two electrons are entangled in such a way that the total spin angular momentum of the pair of electrons is zero (when measured about any axis), if the spin of one electron is measured about some axis and found to be positive along the axis, then the other electron will be found to have spin in the opposite direction along the same axis.

This does not mean that there is a causal connection between measuring the spin of one electron and that of the other. One cannot say one measurement is the cause and the other the effect. This can be illustrated by considering two entangled electrons sent from the center of a long train to its two ends. The spins of these electrons are measured by scientists seated inside the engine and the caboose of the train respectively. Each of these measurements may be called an event. Now the same events are measured from another train going in the opposite direction at a high relative velocity with respect to the first. It will be seen that the temporal order of the events is not the same as measured in the two different trains. This shows that it is not possible to think of one measurement as the cause and the result of the other measurement as the effect, as the order of these events is reversible.

The concept of causality is closely linked with the concept of action at a distance. Since the emergence of Maxwell's theory of electromagnetism in the 19th century, all action is understood as local. Electromagnetic forces are communicated by a field from one charge to another. Likewise gravitational forces are communicated by a gravitational field from one body to another. And these communications take place at a constant speed equal to the speed of light. Entanglement appears to violate this principle of local causality. Influences appear to travel faster than light. But that is not actually the case. Information cannot be sent faster than light from one place to another through entanglement. An alternate way of picturing the same process is by considering time travel of individual particles. Entanglement can be thought of as information being sent backwards in time along the world line of one electron and then forward in time along the world line of the other electron. So there is no action at a distance, and the interaction due to entanglement can be explained as a local interaction, traveling along the world lines.

Chapter 13

Tunneling: Quantum Magic?

13.1 Extending the boundaries of the possible

Quantum theory was born with Planck's hypothesis of the quantization of electromagnetic energy. Energy could not be absorbed or emitted in arbitrary amounts, but only in certain discrete units. The wave nature of an electron meant that only particular electron orbits were permitted in an atom. Angular momentum too was quantized. So it appeared that this new physics introduced some severe restrictions on what was possible.

Heisenberg's Uncertainty Principle added to this sense of restriction. It was impossible to measure both position and momentum with arbitrary accuracy. But the same Uncertainty Principle also opened up certain possibilities that were inaccessible to classical physics. Phenomena that were forbidden by the classical laws of energy and momentum were now possible in certain circumstances.

This is because the Uncertainty Principle made room for the possibility of *virtual processes* that were hitherto forbidden. A virtual process is one that apparently violates the laws of physics, but only temporarily. Because of the Uncertainty Principle, the exact energy of a system could have an uncertainty ΔE within the time interval Δt taken to measure that energy, as long as the two quantities obeyed the Uncertainty Principle:

$$\Delta E \Delta t \gtrsim \frac{\hbar}{2}$$

So the law of conservation of energy could be violated, but only within the duration of the time taken to measure the energy of the system. The total energy of a system could fluctuate as long as the fluctuations were within the limits prescribed by the above Principle. But it must be emphasized that no actual measurement of energy could show a violation of the law of conservation of energy. So if there is a stable system that does not change in time, if we make a measurement and find the energy to be E_1, a subsequent measurement will also yield the same energy E_1. But if a particular process required that the system possessed at least temporarily an energy greater or less than E_1, such a process could take place as long as the initial and final measured energies remained the same.

This highlights an important difference between classical and quantum physics. In classical physics a particle has real physical properties at every instant, whether we choose to measure them or not. In quantum physics, particles do not have well defined physical properties such as position, momentum, energy, etc. in between measurements. One consequence of this feature of quantum theory is that it may be possible to measure the position of a particle at a particular moment of time, and again at a later moment of time, and we really have no way of describing *how* the particle got to the second position from the first.

Of course, all this is simply another way of saying that an electron or a proton or any microscopic object propagates as a wave but is detected as a particle. A measurement is a detection. A measurement will yield particle aspects such as mass, momentum, energy, etc. Between measurements the object has wave-like properties only, which means a total lack of a trajectory. So a quantum system does not have to behave anything like a classical system between actual measurements.

Because the system has the freedom to do forbidden things in between measurements, an electron can get across a barrier that would be impossible were it to obey classical physics. Such a process is called tunneling.

One example of tunneling is a quantum object performing a Houdini escape act by leaking out of a securely locked steel trunk without opening or breaking the trunk.

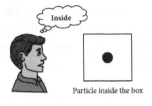

Particle inside the box

Particle outside the box

bsd

Because of tunneling, important physical phenomena such as radioactivity take place which have no explanation in classical physics.

We shall examine the theory of tunneling in the following sections.

13.2 Potential barriers

Let us consider a hollow sphere of radius R made of a conducting metal. Suppose this sphere has been given a positive charge Q coulombs. This charge will distribute itself uniformly over the surface of the sphere. By the laws of electromagnetism, the sphere will acquire a potential V that is related to the charge Q by the formula

$$V = \frac{Q}{4\pi\epsilon_0 R}$$

where ϵ_0 is the constant we encountered earlier, called the *permittivity of free space*.

The laws of electromagnetism also require that the potential at every point on the surface as well as the *space inside* will be the same, equal to $\frac{Q}{4\pi\epsilon_0 R}$ volts. The potential *energy* of an electron in a field having potential V is equal to $-eV$, where $-e$ is the charge of the electron, which is negative.

So an electron moving inside the charged sphere would have a potential energy of

$$-\frac{eQ}{4\pi\epsilon_0 R}$$

This is a negative quantity. And this potential energy would be the same no matter where the electron is as long as it is inside or on the sphere. If the potential energy does not change from point to point that means the charge does not experience a force as long as it is inside or on the sphere. This would be analogous to a smooth frictionless horizontal table top. The gravitational potential energy of a block of wood placed on the surface would be the same at all points on the surface since they are at the same height above the ground. Hence the block would not experience a force in any horizontal direction.

Suppose there is a small opening in the sphere and the electron somehow emerges from the sphere.

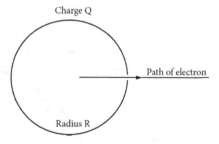

Charge Q

Path of electron

Radius R

The potential energy of the electron at a point outside the sphere drops off as

$$U(r) = -\frac{eQ}{4\pi\epsilon_0 r}$$

where r is the distance of the electron from the center of the sphere. As r increases, $U(r)$ increases from its lowest negative value, and as $r \to \infty$ the potential $U(r)$ increases steadily leveling off at 0 at a long distance from the sphere.

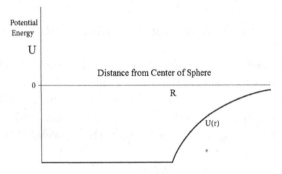

The total energy of a particle is the sum of its potential and kinetic energies:

$$E = T + U.$$

Now, kinetic energy can never be negative. When the electron is inside the sphere its potential energy is negative, and so the total energy may be positive, zero or negative. Suppose the magnitude of the kinetic energy of the electron is less than the magnitude of its potential energy. Then the total energy will be negative. This is the case for the electron shown in the figure below:

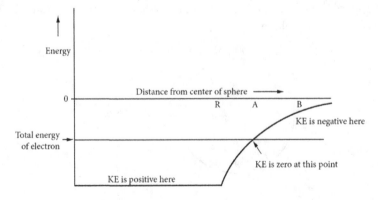

As long as the electron is inside the sphere it has positive kinetic energy. When it emerges from the sphere, the potential energy increases with a consequent decrease in kinetic energy. Expressing the same thing in a different way, as long as the electron is inside the sphere it does not experience any force. But when it emerges from the positively charged sphere it is attracted backwards towards the sphere and therefore its forward motion slows down. This slowing down leads to a decrease of kinetic energy. When the electron reaches the point A it has slowed down to a complete stop. It then retraces its path and goes back into the sphere, like a stone thrown upwards that reaches a high point and then retraces its path downwards.

It is physically impossible for the electron to go beyond A and reach the point B. If it were to do so, it would have negative kinetic energy which is physically impossible because kinetic energy is $\frac{1}{2}mv^2$ and this quantity can never be negative.[1]

So we say that the electron encounters a *potential barrier* when it reaches the point A.

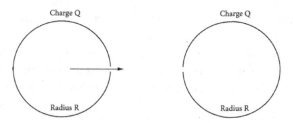

Suppose we now bring a second identical sphere with the same charge with a small opening in line with that of the first, as shown in the figure above. Because of the positive charge on this sphere, if it were brought sufficiently close to the first, an electron emerging from the first would be attracted to the second and so instead of slowing down it would accelerate and enter the second sphere.

But if the second sphere is placed at a sufficient distance from the first, the potential energy of the electron as it emerges from the first sphere would look something like this:

---

[1] This is the formula for a slow moving electron. For an electron moving with a velocity close to that of light, its kinetic energy is given by $mc^2 - m_0 c^2$ where m is the relativistic mass which is always greater than the rest mass m_0. Because m is always greater than m_0, the kinetic energy is always positive.

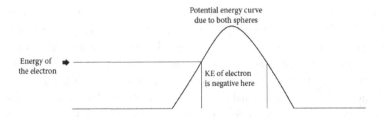

Potential energy curve
due to both spheres

Energy of ➡
the electron

KE of electron
is negative here

This potential curve has the shape of a hill. The horizonal thin line represents the total energy of the electron. While it is inside the first sphere it has a positive kinetic energy. As it emerges from the sphere, its potential energy increases steadily, reaching a maximum halfway between the two spheres, and then decreases as it approaches the second sphere. Classically, it is impossible for the electron to cross this potential barrier. If the electron did reach the portion in the center of the diagram, its potential energy would exceed its total energy and so its kinetic energy would be negative. This is of course not possible. So the electron would come to a stop when its kinetic energy becomes zero, and then retrace its path and return to the inside of the sphere from which it emerged.

The situation is different in quantum theory. Quantum theory allows for a temporary negative kinetic energy as long as no actual measurement is made of the energy. This is the quantum mechanical phenomenon called *tunneling*.

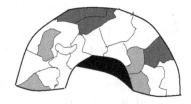

13.3 Tunneling

We shall first illustrate the concept of quantum tunneling with an analogy.

Consider a marble rolling from rest from the crest of a low hill at the point A. By the time it reaches the valley B it has gained sufficient kinetic energy to enable it to roll up the slope of the next hill — which is more like a mountain — till it reaches the point C at which its kinetic energy becomes zero and the marble retraces its path and rolls back down the slope CB.

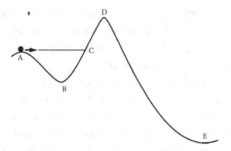

We now switch from physics to a fairy tale. The marble at point A is determined to cross the mountain D and roll down the other side and reach the valley E. However, it knows that it simply does not have the energy to do so. As it rolls up the side of the mountain, it loses all her energy by the time it climbs up to C and has to roll back. But now a fairy comes along and offers it an interest-free loan of energy. (Fairyland is on a tight budget and so fairies cannot afford to give outright gifts.) This fairy offers our marble sufficient energy to climb up the mountain and reach the peak D on the condition that the marble should return this additional energy as soon as it had crossed the peak and entered the downward slope from D to E.

So the marble receives the energy which is equal to the difference in its potential energy between C and D. Having obtained this sudden boost in energy, it clears the peak and then rolls downhill on the other side. It keeps its part of the bargain and returns the temporary gift to the fairy. Thus it would be returning the potential energy difference between C and D.

The net effect of all of this is as though the marble *tunneled* through the mountain from C to a point on the other side at exactly the same height:

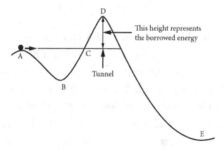

Now instead of a fairy suppose there is a frictionless tunnel cut through the mountain at the height C. It would then be possible for the marble to slide through the tunnel without change of energy and then roll down the slope on the other side of the mountain.

In either case — with help from the fairy or through the tunnel — the marble would reach the valley at E with exactly the same kinetic energy.

This gravitational problem is an analogy to the electrostatic problem of the electron emerging from one charged sphere and having insufficient energy to overcome the high potential barrier in order to enter the attractive field due to the second charged sphere. While this is impossible according to classical physics, it may be possible at the quantum level. If the potential barrier is not too high, and not too wide, it would be possible for the electron to "borrow" some energy from the surrounding electric field and return the energy after the electron has cleared the hurdle. So the fairy who lends energy at zero interest is quite real at the quantum level! In quantum physics we do not describe this process as borrowing and returning but as "tunneling".

13.4   Tunneling and alpha decay

The theory of quantum mechanical tunneling was used to explain the phenomenon of radioactivity. A Uranium 238 nucleus is not very stable. When left to itself it changes to a Thorium 234 nucleus by emitting a Helium 4 nucleus. The Uranium nucleus contains 92 protons and 146 neutrons. A Thorium nucleus contains 90 protons and 144 neutrons. A Helium 4 nucleus contains two protons and two neutrons. The nuclear reaction can be written as

$$U_{92}^{238} \rightarrow Th_{90}^{234} + He_2^4$$

A Helium 4 nucleus is also called an alpha (α) particle. So a Uranium 238 nucleus can be thought of as a combination of a Thorium 234 nucleus and an alpha particle. More specifically, we can think of this combination as an alpha particle trapped inside a Thorium 234 nucleus. This α particle is free to move within the Thorium nucleus, but cannot escape because there is a potential barrier in the shape of a hill at the boundary or surface of the Thorium nucleus. At this boundary the kinetic energy of the α particle becomes zero and beyond that it becomes negative for a short distance.

This potential barrier is due to the very strongly attracting nuclear force that acts between the particles inside the nucleus — viz. the protons and the neutrons. Though two protons would repel each other because of their like positive charges, when they come sufficiently close to each other the strong nuclear force of attraction overcomes the electric force of repulsion. But when these particles move sufficiently far apart from each other, the electric force becomes stronger than the nuclear force, and so they begin to repel each other. So the alpha particle is strongly attracted to the Thorium nucleus by the nuclear force, but if it could clear the potential barrier then it would reach a place where the electric force of repulsion becomes stronger than the attractive nuclear force, enabling the alpha particle to flee.

Classically, it would be impossible for the α particle to escape, but quantum theory allows the particle to tunnel through the potential barrier and emerge from the nucleus. Once the alpha particle emerges from the nucleus the electrostatic forces will become predominant. The alpha particle is positively charged, and so is the residual Thorium nucleus. Hence the newly liberated alpha particle will be repelled by the Thorium nucleus and thus fly away. This is called alpha radioactivity.

The probability for such an event to occur is very low, and so alpha decay takes place extremely slowly. If we have a sample of 1 kg of Uranium 238, it would take about 4.5 billion years before the amount of Uranium had decreased to 0.5 kg as a result of alpha radioactive decay. This length of time is called the *half life* of U^{238}.

There are other kinds of radioactivity, some more harmful to human life than others. Nuclei also emit electrons, which were originally called β (beta) particles, before their identity became known, and also photons of very high frequency called γ (gamma) rays. The study of radioactivity in general belongs to nuclear physics.

Radioactivity is an irreversible process. According to the Second Law of Thermodynamics radioactive decay contributes to the increase of entropy of the universe.

13.5 Summary

According to classical mechanics, every particle has a definite set of real properties such as mass, velocity, momentum, kinetic energy, etc. at all times. When a particle is moving in force field such as a gravitational or electromagnetic field, its total energy is always constant. The total energy is the sum of the potential and kinetic energies. Now, the potential energy can be positive or negative, but the kinetic energy can never be negative. When a particle is in a bound state, such as an electron orbiting a nucleus, its total energy is negative because the negative potential energy has a greater magnitude than the positive kinetic energy. This electron is bound to the nucleus by the force of attraction between itself and the nucleus, and cannot break free — unless it receives energy from the outside. But there may be a situation where a particle is confined by a binding force which acts only for a short distance, and this force disappears at longer distances. So here even though classical physics will not permit the particle to escape, quantum physics allows the particle to escape through a syphonic effect called tunneling.

Consider a negatively charged particle moving freely inside a positively charged hollow sphere. Such a particle would feel no force as long as it is inside the sphere. And if there is a small opening on the sphere, the charged particle — even if it could emerge through the opening — will not be able to escape from the sphere because the force of attraction between the sphere and the particle will pull the particle back inside. But if another identical hollow sphere which is also positively charged with a similar opening is brought close to the opening of the first sphere, then the particle — if it has sufficient kinetic energy — could escape from the first sphere and enter inside the second sphere.

Now, according to quantum theory, even if the charged particle did not have sufficient kinetic energy to escape from the first sphere and enter the second sphere, there is some possibility that it could indeed escape the attractive force of the first sphere and enter the second sphere. This is because of the uncertainty principle.

As long as we do not measure the energy of the particle at all times, it is possible for the particle to move through a zone which would be forbidden by classical physics. Classical mechanics would deem that it is impossible for the particle to have negative kinetic energy. In quantum theory it is indeed impossible *to observe* a particle having negative kinetic energy, but a temporary negative kinetic energy is possible as long as it is not observed.

So the particle travels through a forbidden zone called a potential wall through a process called tunneling. Tunneling is responsible for forms of radioactivity.

Chapter 14

The Spatial Wave Function

14.1 Probability density

In this book we have largely dealt with particles inside a cavity or a chamber within a cavity. We have asked questions such as the probability of finding a particle in one or the other chamber. But often our interest is in locating a particle more precisely than just a large chamber. But since a particle does not have an exact trajectory in quantum theory, rather than talking about finding the particle at a particular geometric point, we should talk about the probability of finding the particle within a small region of volume which contains the point. A photographic plate records the spot at which a photon hits the plate. But this spot is not a geometric point. If it were, it would not be visible, and we could never detect the photon. The spot is actually a smudge, smeared over a small area. So we cannot localize a photon beyond a small area. Actually, the spot on the photographic plate also has a small depth. So we have really localized the photon to within a small volume, and not a geometrical point. So a realistic question could be: What is the probability of finding the particle within a small volume ΔV? It is this problem that we shall discuss in this section.

Since ΔV is very small, we shall call it a *volume element*. We first define a quantity that we call a probability density ρ which has the meaning that the product $\rho \Delta V$ is the probability that the particle is found within the volume element ΔV. Let us choose a point somewhere within this volume element and label this point by its coordinates as $P(x, y, z)$. Since we expect that the probability density varies from point to point, it is in general a function of the coordinates of the point: $\rho(x, y, z)$. Since ρ is a scalar and in general varies from point to point, we call it a *scalar point function*.

If we know for sure that the particle is somewhere inside a large box, the probability of finding the particle inside the box is 1. If we divide up the box into a large number of small volume elements centered at different points within the box, and the volume element localized at the point which we label as i is called ΔV_i, then the probability of finding the particle within the very small volume element ΔV_i is equal to $\rho(x_i, y_i, z_i)\Delta V_i$. Of course, the label i has nothing to do with the imaginary number unit $\sqrt{-1}$. It is a label meaning "any number" or "some number" which can have a value from 1 to a very large number equal to the number of volume elements into which we have divided up the space under consideration. Clearly, the probabilities of finding the particle in the different volume elements must all add up to 1.

For example, we could imagine a box of volume 2 cubic meters which has been divided into 5×10^{10} small volume elements each of volume $\Delta V_i = 4 \times 10^{-11}$ cubic meter. So the subscript i can take on all the values from 1 to 5×10^{10}. If the probability of finding the particle inside the ith volume element is P_i, then

$$P_1 + P_2 + P_3 \ldots + P_{5 \times 10^{10}} = 1$$

We could write the above equation also as $\Sigma_i P_i = 1$ where the symbol Σ_i is shorthand for summing up over the terms having all the possible values that i can take.

14.2 Amplitude and probability

We saw earlier (Ch. 10) that a probability P for an event to occur — such as detecting a photon on one of the walls of the box — is related to the amplitude α for the event to occur by the equation

$$P = \alpha^* \alpha = |\alpha|^2$$

We make a slightly different definition when we come to spatial probability, or the probability of finding the particle in some small volume located at a point.

We define an amplitude ψ, which we call the *wave function*, which is related to the probability density of finding the particle at a point i by the equation

$$\rho(x_i, y_i, z_i) = \psi^* \psi = |\psi|^2$$

and so the probability of finding the particle inside a small volume element ΔV_i is given by

$$P_i = \psi^*(x_i, y_i, z_i)\psi(x_i, y_i, z_i)\Delta V_i$$

where (x_i, y_i, z_i) are the coordinates of a point within the volume element ΔV_i.

If we know for certain that the particle is somewhere within a fixed space, we can divide up the space into a large number of tiny volume elements, and if we add up the probabilities of finding the particle in each of the volume elements we should get 1 as the total:

$$\Sigma_i P_i = \Sigma_i \psi^*(x_i, y_i, z_i)\psi(x_i, y_i, z_i)\Delta V_i = 1$$

So if we have divided up the region into 50 billion tiny volume elements, then i would range all the way from 1 to 50 billion. Such sums — which are basically what we call *integrals* in calculus — are carried out by computers using suitable codes. Of course, most of the time we do not require so many tiny volume elements. The volume can be divided into fewer and larger pieces and still provide an accurate value of the integral. In some simple cases there are ways of obtaining the values of the integrals using formulas which can be derived using the rules of calculus.

14.3   The wave function and measurements

Knowing the wave function enables us to calculate the probability of finding a particle in some region in space. But the wave function provides us with much more information. So physical quantities such as the momentum, the energy, the angular momentum, etc. of the particle can also be found from the wave function. In order to obtain these quantities, we need to construct *operators* which will work on the wave function. Earlier (Ch. 10) we came across operators that operated on a state function such as $|0\rangle$. The operators we are now seeking operate on the wave function ψ.

The operator is a mathematical entity that works on a function. A trivial example of an operator is the number 2. When this operator acts on a function ψ, it has the effect of converting the function ψ into another function, viz. 2ψ, which has twice the magnitude of ψ. The number 2 is constant, and so this operator will not reveal to us anything that we do not already know.

But suppose we want to know where the particle is. If we were to measure the position of a particle in some laboratory setting, we might get a particular value. So if we are measuring the x position of a particle, we might get the value a_1 when we perform a measurement. If we repeat the experiment with identical conditions as before, we may not get exactly a_1, but a different value that we shall call a_2. If this experiment is repeated several times, each time keeping the initial conditions of the experiment the same as the previous time, then we would get a number of different values for the x position, such as a_1, a_2, a_3, etc. The average of a large number of such independent measurements is called *the expected value of the position*, written as $\langle x \rangle$. So

$$\langle x \rangle = \frac{a_1 + a_2 + a_3 ... a_N}{N}$$

Quantum mechanics is primarily about calculating the expected value of a physical quantity using the wave function. Since $\rho \Delta V$ is the probability of finding the particle within the region ΔV, the expected value of the x position of the particle is simply

$$\langle x \rangle = \Sigma_i^N x_i \rho(x_i, y_i, z_i) \Delta V_i$$

where the summation symbol Σ_i^N indicates that the product has to be summed over all the values of i from 1 to N where N is the large number of volume elements into which the region has been divided. Since $\rho = \psi^* \psi$, we get the expected value of the x position from the equation

$$\langle x \rangle = \Sigma_i^N x_i \psi^*(x_i, y_i, z_i) \psi(x_i, y_i, z_i) \Delta V_i \qquad (14.1)$$

Here it is understood that the ψ functions should satisfy the equation

$$\Sigma_i^N \psi^*(x_i, y_i, z_i) \psi(x_i, y_i, z_i) \Delta V_i = 1 \qquad (14.2)$$

which implies that the probability of finding the particle somewhere within the region is 1. Equation (14.2) is called *the normalization condition*.

As the number N of such volume elements becomes very large, and the size of each volume element ΔV becomes very small, Eq. (14.1) can be written in the symbolic form of calculus as

$$\langle x \rangle = \int_V x \psi^* \psi dV$$

where the letter V appended to the integral symbol $\int$ means the integration — or summation — is over the entire region which we label by the symbol V. The normalization condition can be written as

$$\int_V \psi^* \psi dV = 1$$

What about the expected value of the momentum of a particle? This is a somewhat more complex affair. Whereas the position operator x is identical with the position x, the momentum operator depends on the wave function at more than one position. At the risk of oversimplification, we could say that the momentum operator measures the rate at which the wave function varies with distance. Without going into the derivation, we state that the momentum operator $\hat{p}_x$ is defined by the equation

$$\hat{p}_x\psi = -i\hbar\frac{\Delta\psi}{\Delta x}$$

or to use the symbolism of calculus:

$$\hat{p}_x\psi = -i\hbar\frac{\partial\psi}{\partial x}$$

And so

$$\langle p_x \rangle = -i\hbar \int_V \psi^* \frac{\partial\psi}{\partial x} dV$$

The momentum operator itself can be written as

$$\hat{p}_x = -i\hbar\frac{\partial}{\partial x}$$

This momentum operator is applicable to particles traveling at all speeds — for photons traveling at the speed of light and for electrons traveling slowly or very fast. Using the momentum operator it is possible to construct a kinetic energy operator.[1] Suppose we wish to know the kinetic energy of an electron. In classical physics we would seek to find the momentum or the velocity and use the formula to find the kinetic energy. In quantum physics we use the kinetic energy operator, which we shall denote by the symbol $\hat{T}$, and apply it to the wave function to obtain the kinetic energy. Now, since every prediction in quantum mechanics is statistical, our theoretical result would equal the average kinetic energy of the electron inside the box. This result is called the *expected value* of the kinetic energy. As explained earlier, the expected value is not really what we would expect to get with a single experimental observation. The expected value is the *average* of the values we would get through repeated observations on the same system. We could imagine a million identical boxes each with a free electron inside the box, we measure the kinetic energy of the electron in each box, and take the average of all the observations. This statistical result can be obtained either through a very large number of observations, which we call an *a posteriori* result, or we could calculate it mathematically

---

[1]The kinetic energy ($p^2/2m$) operator for a particle moving much slower than light is written in calculus notation as

$$-\frac{\hbar^2}{2m}\left(\frac{\partial^2}{\partial x^2} + \frac{\partial^2}{\partial y^2} + \frac{\partial^2}{\partial z^2}\right)$$

without doing any experiment, which we call an *a priori* result. The success of quantum mechanics is shown by the fact that the theoretical result agrees excellently with the experimental result in an overwhelmingly large number of situations.

The *a priori* expected value of the kinetic energy — written as $\langle \hat{T} \rangle$ — of an electron can be calculated from the wave function as follows:

$$\langle \hat{T} \rangle = \Sigma_i \psi^*(x_i, y_i, z_i) \hat{T} \psi(x_i, y_i, z_i) \Delta V_i \tag{14.3}$$

The expected value of the kinetic energy of the electron is obtained by taking the average value of the kinetic energy of the electron over the entire region where the electron has some probability of being found. Each volume element ΔV_i provides a contribution to the kinetic energy, and the total kinetic energy is the sum of all these contributions. In Eq. (14.3) we have a sum over a large number of terms. Each term (with the subscript i) is equal to the value of the kinetic energy of the electron *if the electron is found inside the element* ΔV_i multiplied by the probability of finding the electron inside ΔV_i.

This sum of Eq. (14.3) is written in calculus notation as

$$\langle \hat{T} \rangle = \int_V \psi^*(x, y, z) \hat{T} \psi(x, y, z) dV$$

What about a potential energy operator?

Now, potential energy varies in general from one point to another. If we represent the potential energy of an electron by the letter U, then U is a function of the spatial coordinates (x, y, z) and so we write $U(x, y, z)$. An example of such a function would be $U(x, y, z) = 3xy + 4y^2 z^3$. Suppose we ask the question: what is the potential energy of the electron inside a volume where the potential energy is a function of the spatial coordinates?

If we knew for sure that the electron is at the point having the coordinates $(3, -1, 2)$, we could find the potential energy by substituting the values of x, y and z into the formula for potential energy. In the example just given this quantity becomes $U(3, -1, 2) = 3 \times 3(-1) + 4(-1)^2 2^3 = 23$ joules. However, we do not know exactly where the electron would be found if we searched for it. We only have a wave function that tells us the probability of finding the electron somewhere. So we multiply the potential energy calculated within some volume element by the probability of finding the particle within that element. This product gives the "contribution" to the potential energy of the particle from the wave function within that volume element. We add up the products for all the volume elements of the entire region within which the particle is confined to obtain the total potential energy. So each volume element provides a small contribution to the potential energy. Those elements with greater probability density will make greater contributions.

So, the expected value of the potential energy of a particle in a region where the potential energy is given by the function $U(x, y, z)$ is obtained by the sum

$$\Sigma_i U(x_i, y_i, z_i)\psi^*(x_i, y_i, z_i)\psi(x_i, y_i, z_i)\Delta V_i$$
$$= \Sigma_i \psi^*(x_i, y_i, z_i)U(x_i, y_i, z_i)\psi(x_i, y_i, z_i)\Delta V_i$$

It is conventional to express the potential energy function as an operator acting on the function ψ even though the potential energy is just a factor and not an operator like the momentum or kinetic energy operators. This sum is written in calculus notation as

$$\langle U \rangle = \int_V \psi^*(x, y, z)U(x, y, z)\psi(x, y, z)dV$$

Thus we have obtained a quantum mechanical formula for finding the potential energy of an electron from its wave function and the expression for the potential energy. Now, if we know the geometry of the system, we could find an expression for the potential energy. For example, if we have an electron approaching a large uniformly charged plate, its potential energy is given by kx where x is the perpendicular distance of the electron from the plate and k is a constant that depends on the surface charge density on the plate and the charge of the electron as well as the constant ϵ_0. The potential energy of an electron approaching another negative electron is obtained from Coulomb's Law and works out to

$$\frac{e^2}{4\pi\epsilon_0 r}$$

where r is the distance between the two electrons.

If we denote the kinetic energy operator by the abbreviated symbol $\hat{T}$ then we can define the total energy operator H called the *Hamiltonian* as $H = \hat{T} + U$.

For an electron in an atom or in some other system which has a constant total energy E (as distinct from an electron colliding with another particle and thereby losing or gaining energy) we may write

$$H\psi \equiv \hat{T}\psi + U\psi = E\psi$$

This is a simple form of the basic equation used in quantum mechanics called the *Schrödinger Wave Equation*.[2]

---

[2]In calculus notation the Schrödinger Wave Equation for a particle with constant energy E is written as $-\frac{\hbar^2}{2m}\left(\frac{\partial^2}{\partial x^2} + \frac{\partial^2}{\partial y^2} + \frac{\partial^2}{\partial z^2}\right)\psi + U(x, y, z)\psi = E\psi$.

It is possible to obtain operator expressions for physical quantities of interest to us such as the momentum, the kinetic energy, the total energy, the angular momentum, etc. The precise shape of these operators is expressed in the symbolic language of calculus. In quantum theory physical variables are obtained from the system by applying a mathematical operator to the wave function ψ.

In some situations the energy of the particle is not constant in time. In such a case ψ is also a function of time, as well as x, y and z. The Schrödinger wave equation would then contain a term which expresses the rate at which ψ changes with time.

14.4   A historical note

Historically the wave function ψ was introduced to explain the wave aspect of particles such as electrons. Schrödinger created his wave equation as a mathematical description of particles that had a wavelength according to de Broglie's law. He solved his wave equation and obtained the values of ψ in different situations, such as an electron orbiting a proton in the hydrogen atom. The solution of the wave equation yielded the correct energy levels of the hydrogen atom and successfully explained the frequencies of the radiation emitted or absorbed by the hydrogen atom, which is the simplest of the atoms. While the Schrödinger wave equation was therefore a successful tool in quantum mechanics, the precise meaning of the wave function itself was not very clear. It took some time for the evolution of the interpretation offered in this chapter. This interpretation is called the statistical interpretation and is a central feature of the mainstream understanding of quantum theory today. The statistical interpretation was first suggested for photons by Einstein and later developed systematically and applied to electrons and other particles by Max Born. The wave function and its statistical interpretation spelt the death knell of the determinacy of Newtonian physics. One cannot predict the exact future; one can only assign probabilities for specific outcomes. And in order to calculate the probabilities, we require the wave function which is obtained as a solution of Schrödinger's equation.

14.5   Summary

Every particle can be represented by a spatial wave function written as ψ which is a complex function of the spatial coordinates (x, y, z). The probability of finding the particle within a small volume ΔV is given by $\psi^*\psi\Delta V$. If we know the wave function we can calculate the *expected values* of the position, the momentum, the kinetic or the potential energy of the particle. In order to do that, we need the operators corresponding to these physical quantities. The operator for position is simply x (or y or z). The operator for momentum is $-i\hbar\frac{\partial}{\partial x}$. Using these two operators we can construct operators for potential energy, kinetic energy, angular momentum, etc. In general, if we want to find the expected value of some quantity A, and if the operator corresponding to that quantity is $\hat{A}$, then the expected value of this quantity is given by $\langle A \rangle = \int_V \psi^* \hat{A}\psi dV$.

Exercise 14.1. A particle is present inside a cubical box of length L with non-absorbing walls. We choose a coordinate system with the origin at one corner of the cube and the coordinate axes along three sides. One possible wave function of a particle in such a container is

$$\psi = \sqrt{\frac{\pi^3}{8L^3}} \sin\frac{\pi x}{L} \sin\frac{\pi y}{L} \sin\frac{\pi z}{L}$$

This particular wave function is entirely real, and so $\rho = \psi^*\psi = \psi^2$. To simplify our work let us assume that the length of the box L equals π meters, so that the wave function takes on the simpler form $\psi = \sqrt{\frac{1}{8}} \sin x \sin y \sin z$. The coordinates of the corners of the box are now $(0,0,0), (0,\pi,0), (0,0,\pi), (\pi,0,0), (\pi,\pi,0), (\pi,0,\pi), (0,\pi,\pi), (\pi,\pi,\pi)$. Find a calculator that can give you the trigonometric functions and set the angle to radian mode.

1. Show that the wave function is zero at all 8 corners.

2. Show that the wave function is zero at all the faces (for which at least one of the three coordinates is either 0 or π).

3. Show that the value of ρ at the center of the cube $(\pi/2, \pi/2, \pi/2)$ is $\frac{1}{8}$.

4. Does ρ have the highest value at the center? To check if this is so, try changing the values of x, y and z slightly, say by $+0.1$ or -0.1 and find the value of ρ for each value. If ρ has a maximum value at the center, the value of ρ would decrease when we move to any of the neighboring points.

Chapter 15

Conclusion

15.1 Summary

15.1.1 *Wave and particle*

And so, we have learned the basic concepts of quantum theory. Without actually delving into the mathematical subtleties of this theory we found that quantum theory is a mysterious but coherent way of understanding physical reality.

The very first thing we learned is that electromagnetic energy is quantized. It can be absorbed or emitted only in discrete units, and each unit has a value proportional to the frequency of the radiation:

$$\epsilon = h\nu$$

This one equation captures the essence of quantum theory. Electromagnetic energy — and light in particular — is not continuous, but discrete. Each discrete unit or photon is a material particle with mass, momentum and energy. But these very material attributes of the photon are connected inextricably with the wave nature of electromagnetic radiation:

$$\text{energy} = h\nu$$

$$\text{momentum} = \frac{h\nu}{c}$$

$$\text{mass} = \frac{h\nu}{c^2}$$

Each one of these material attributes of a photon is dependent on its wave nature, because *frequency* is entirely a wave property.

Every particle having momentum p has a de Broglie wavelength given by $\lambda = h/p$. The only allowed orbits of an electron around a nucleus are

179

those for which the circumference is an integer multiple of the de Broglie wavelength of the electron. Thus the energy levels of simple atoms can be calculated.

The wave nature of the electron was explicitly confirmed by its ability to undergo diffraction and interference, phenomena which have no place in an exclusively corpuscular understanding of electrons.

15.1.2 *A statistical result*

Interference of light waves happen even when only one photon is present, but this phenomenon can be observed as such only when a very large number of photons are sent through the apparatus. The interference pattern generated by two closely spaced slits is the result of an extremely large number of photons impinging upon the detection screen. Thus the wave nature of light is demonstrated statistically. It cannot be demonstrated with only one photon. It requires either a very large number of independent observations involving a single photon each time, or a single observation involving a very large number of photons at the same time. Thus the wave nature of light is something that is established statistically.

Everything that was said in this foregoing paragraph applies to electrons and all other "material" particles.

Statistics is an unavoidable aspect of quantum theory, because there is a built-in freedom in nature. There is no law that states that a particle such as a photon or an electron *must* be found in a particular place or with a particular momentum. One can only talk about the probability of occurrence of a particular event.

15.1.3 *Uncertainty principle*

There is no determinism in quantum theory. One cannot predict the exact outcome, but only the probability of occurrence of a particular outcome. In classical physics the motion of a particle can be described by a trajectory. This trajectory can be calculated beforehand and one can predict where the particle will be at any time in the future. There are no trajectories in quantum theory.

There are no trajectories because it is impossible to measure both the position and the momentum in any direction with arbitrary accuracy. If we try to measure both, we end up with unavoidable uncertainties in both. If the uncertainty in position is Δx and that in momentum is Δp_x then

Heisenberg's Uncertainty Principle states that

$$\Delta x \Delta p_x \gtrsim \frac{\hbar}{2}$$

15.1.4 *Wave functions and operators*

Any kind of measurement involves an interaction with the entire particle or quantum, not just one aspect of it. A particle is absorbed and/or emitted in every interaction. The absorption of a particle is represented by an annihilation operator a and the emission of a particle is represented by a creation operator $a^\dagger$. These operators act upon field states that express the number of particles in the field.

Intervals of length and intervals of time are not absolute but are relative, meaning that a total description of a process involves both length and time. Mathematically, the coupling between space and time is expressed by imaginary numbers. Because every interaction in quantum physics involves the whole particle, and therefore not just the spatial dimensions or just the time dimension, quantum measurement involves a coupling of space and time. So in quantum mechanics the operators that represent these interactions are in general composed of not just real but also imaginary numbers. Thus complex numbers are indispensable in quantum mechanics.

15.1.5 *Spin and statistics*

Every particle has intrinsic spin, expressed by a spin quantum number which can be 0 or any whole multiple of $\frac{1}{2}$. Particles with spins that are odd multiples of $\frac{1}{2}$ are called fermions. They obey Pauli's Exclusion Principle, whereby no two identical particles can be in exactly the same quantum state, i.e. have the same set of quantum numbers. Fermions obey Fermi–Dirac statistics. Particles with spins that are even multiples of $\frac{1}{2}$ are called bosons. Any number of identical bosons can occupy the same quantum state. Bosons obey Bose–Einstein statistics.

15.1.6 *Zitterbewegung*

Upon incorporating the Theory of Relativity into the quantum theory of electrons it was found that an electron can be in a positive or a negative energy state. Dirac's equation shows that an electron constantly switches between positive and negative energy states, a process called *zitterbewegung*. In this process, the electron is also continually displaced at the speed of light *c* back and forth between two very closely spaced points. It is meaningless to talk about the trajectory of the electron between these points.

15.1.7 *Antiparticles*

Every particle has a corresponding antiparticle. The photon is its own antiparticle. The positron is the antiparticle of the electron. The positron can be thought of as an electron traveling backwards in time. The positron can also be thought of as a hole in a sea of negative energy electrons. Either way, the physical behavior of the positron is the same. An electron and a positron would annihilate each other to produce a pair of photons. The antiproton is the antiparticle of the proton. Even neutral particles have their antiparticle twins. So there are antineutrons and antineutrinos. It is theoretically possible for a galaxy to be formed exclusively from the antiparticles of our familiar electrons, protons and neutrons. Such a galaxy would be made of antimatter. Both matter and antimatter have positive mass.

15.1.8 *Entanglement*

Two or more particles can be entangled, i.e. they can be prepared jointly in a single quantum state. This quantum state is preserved even when the particles are separated by a large distance. One way of entangling two electrons is to prepare them in such a way that their total spin angular momentum is zero. So a measurement of the angular momentum of one electron along any axis will reveal the angular momentum of the other electron along the same axis no matter how far away the electrons are. But this does not permit the communication of information from one place to another at arbitrarily high speeds. Thus even in entanglement there is no action at a distance that violates the Special Theory of Relativity.

15.1.9 *Tunneling*

A particle can cross a potential barrier as long as this barrier is not too high and not too wide. The kinetic energy of the particle becomes negative within this barrier but since no measurement is made on the particle within this classically forbidden zone the particle can pass through this barrier. This phenomenon called tunneling can be understood by the uncertainty principle. Tunneling is responsible for alpha radioactive decay.

15.2   The next step

This book is a introductory guide to quantum theory. It is not meant to be a complete textbook, but it has covered the basic concepts of the subject. You have therefore learned the essentials of this important area of modern physics. If you feel that you do not understand the subject, you are in good company. The greatest quantum physicists all agree that nobody understands quantum theory. Quantum theory is a mystery, and the only valid approach to this mystery is to contemplate it. The purpose of this book is to introduce you to this mystery so that you may contemplate it for the remainder of your life.

Of course, one can learn to solve problems, and knowing which methods to use in which situations is the task of quantum mechanics. That is the practical aspect of knowing quantum theory, which for most physicists is the only aspect that counts, expressed as the "Shut up and calculate!" principle. Most calculations of quantum mechanics are carried out using wave functions and wave equations.

You have acquired some familiarity with electrons and photons and have learned the names of the neutron and the proton. Numerous other particles have been discovered besides the electron, proton, neutron and the photon. The study of these particles is an important and interesting branch of physics. Some introductory texts on quantum theory discuss the basic theories of elementary particles. But since a knowledge of elementary particle physics is not required to obtain a basic understanding of quantum theory, we have all but omitted this subject from our book.

There are excellent monographs on quantum theory that can pick you up from where you are and transport you to a higher level. The bibliography provided at the end of this book is but an incomplete selection of the breathtaking assortment of textbooks available in the market. A very brief explanation is provided as an aid to navigating through this bewildering list of titles on this rich and colorful subject.

Finally, in this book we have avoided all philosophical discussions on quantum theory, of which much has been written. Because of the abstract nature of quantum theory, not only philosophers, but also theologians, psychologists, spiritualists, artists and musicians have all felt they have something to say about this discipline. A small selection of books of this nature is included at the end of the bibliography.

Appendix A

Answers to Exercises

Most of the answers have been rounded to two significant figures.

2.1.

(a) 2.6×10^6 s

(b) 3.2×10^7 s

2.2.

(a) 2500 N

(b) 3.0×10^4 m/s

(c) 6.0×10^{-3} m/s^2

2.3.

(a) 2.0×10^{30} kg

(b) 1.6 m/s^2

(c) 7.9×10^3 m/s

2.4.

(a) 2.7×10^{33} J

(b) 1.2×10^7 J/s or W (watts)

3.1. (a) 0^0 F (b) -20^0 F (c) Same (d) -60^0 C

3.2.

(a) 3.3×10^{22}

(b) 98 g/mol

4.1. 1.2×10^{-12} N

4.2.

(a) 3000 N

(b) 750 m/s^2

6.1

Write two equations for the different frequencies and kinetic energies:

$$h\nu_1 = T_1 + \phi$$
$$h\nu_2 = T_2 + \phi$$

So $h(\nu_2 - \nu_1) = T_2 - T_1$

Hence $h = \frac{T_2 - T_1}{\nu_2 - \nu_1} = \frac{(1.96-1.3)\times 10^{-19}}{1\times 10^{14}} = 6.6 \times 10^{-34}$ (SI units)

6.2

Momentum $= 2.8 \times 10^{-27}$ kg m/s

Mass $= 9.4 \times 10^{-36}$ kg

6.3

Wavelength $= 2.4 \times 10^{-12}$ m

6.4

5.76 J

6.5

4.4×10^{-23} kg m/s

7.2

(a) 5.0×10^{-11} m

(b) 1.2×10^{-11} m

8.1

Relative Speed $= 0.99995c$

8.2.

1 (a) 4.6×10^{-19} J

1(b) 5.1×10^{-36} kg

8.3.

1. $0.87c$

2. $0.93c$

3. 3.8×10^{-12} J

4. 21.3 minutes

9.1

Time-like interval

9.2

(a) Space-like interval

(b) No. In order for one event to have an effect upon the other it should be possible to send at least a signal from one event to the other event before this latter event occurred. Since the time interval between the two events is less than the time taken for light to travel from one place to the other, it is impossible for one event to have an effect on the other.

10.1

1. (a) 5.00 (b) 10.00 (c) 2.83 (d) 0.13

2. (a) $22 + 7i$ (b) $22 + 7i$ (c) 6.4 (d) 3.6 (e) $2 + 23i$ (f) $2 - 23i$

10.2

(A) 0 (B) 0 (C) 1 (D) 0

11.1.

1. (a) $\sqrt{2}\hbar$

(b) $-\hbar, 0, \hbar$

2. (a) $\sqrt{12}\hbar$

(b) $-3\hbar, -2\hbar, -\hbar, 0, \hbar, 2\hbar, 3\hbar$

11.2.

45^0, 90^0 and 135^0

Appendix B

Bibliography

Suggestions for Further Reading

Now that you have made an entry into quantum theory and learned the basic rules of quantum mechanics you are now poised to increase your knowledge of the subject. To begin with, in order to appreciate the sheer oddness of quantum theory, it helps to know how this theory developed. While we have provided the barest outline in this book, there are more detailed treatments available, some written by the pioneers of the subject. Some of these histories also discuss the concepts of quantum theory. But there are other books that focus on the concepts themselves. Both sorts of books would help you obtain a better conceptual grounding in the subject. Some of you who have completed the present book may wish to learn more about the subject and read books explaining the mathematical methods of quantum mechanics. At another level, some of you may like to know how quantum theory has influenced disciplines outside of physics such as philosophy, theology, music, etc.

The following bibliography offers a representative collection of works that cover the areas described above.

Historical

J. P. McEvoy and Oscar Zarate, *Introducing Quantum Theory: A Graphic Guide to Science's Most Puzzling Discovery* (Icon Books, 2003). A very basic historical outline with copious cartoon style illustrations of the physicists who built quantum theory. The book is an entertaining summary of the historical development of the subject.

Louis de Broglie, *The Revolution in Physics: A Non-mathematical Survey of Quanta* (Original 1953; Greenwood, 1970). A firsthand narration of the emergence of quantum theory by one of its greatest pioneers and architects.

Banesh Hoffmann, *The Strange Story of the Quantum* (Original 1959; Dover, 2011). A lucid non-technical classic of the history of quantum theory.

George Gamow, *Thirty Years that Shook Physics* (Doubleday, 1966; Dover reprint 1985). The classic history of the birth of quantum theory by an outstanding physicist and master science writer who himself contributed significantly to that history.

Barbara Lovett Cline, *Men Who Made a New Physics: Physicists and the Quantum Theory* (University of Chicago, 1987). A history of quantum theory with closer attention given to the physicists and their relationships.

Louisa Guilder, *The Age of Entanglement* (Vintage, 2009). The story of quantum entanglement, told in a witty and charming style that is faithful to the historical facts.

The following historical works also provide important insights into concepts:

John Gribbin, *In Search of Schrödinger's Cat: Quantum Physics and Reality* (Bantam, 1984). A very readable account of the historical and conceptual development of quantum theory by a skilled popularizer of science.

Victor Guillemin, *The Story of Quantum Mechanics* (Scribner, 1968). A comprehensive history of the concepts of quantum theory with a substantial exposition on elementary particles. It also contains a section on the philosophical implications of quantum theory.

Conceptual

John Polkinghorne, *Quantum Theory: A Very Short Introduction* (Oxford University, 2002). This excellent historical and conceptual summary uses no mathematics except in one section where the author feels it is unavoidable for understanding the matter.

Michael D. Fayer, *Absolutely Small: How Quantum Theory Explains Our Everyday World* (AMACOM, 2010). A detailed but non-mathematical description of the concepts of quantum theory.

Alastair Rae, *Quantum Physics: A Beginner's Guide* (Oneworld Publications, 2005). A lucid and non-mathematical introduction to the subject.

George Gamow, *One, Two, Three, Infinity* (Viking, 1961; Dover reprint, 1988). Possibly the greatest popular work on modern physics ever written, this book may have played a greater role in launching young readers into a career of physics than any other publication.

George Gamow, *Mr. Tompkins in Paperback — Mr. Tompkins in Wonderland (1940) and Mr. Tompkins Explores the Atom (1944) published as a single volume* (Cambridge, 2012). Delightful whimsical fantasies that expound solid scientific principles of quantum theory and relativity, amply illustrated with humorous sketches.

Marcus Chown, *Quantum Theory Cannot Hurt You* (Faber, 2008). Popular non-technical exposition of the subject which also offers a basic introduction to the concepts of Relativity.

Valerio Scarani, *Quantum Physics: A First Encounter* (Oxford, 2006). This discussion of important concepts of quantum theory from a modern twenty first century perspective is very accessible and helpful.

David Z. Albert, *Quantum Mechanics and Experience* (Harvard, 1993). Despite the title this is a purely conceptual exposition of the basic principles of quantum theory. This book will appeal to readers who have a taste for abstract, logical reasoning.

Art Hobson, *Tales of the Quantum: Understanding Physics' Most Fundamental Theory* (Oxford, 2017). An excellent treatment of quantum theory for beginners. The author explores the concept of a quantum, the basic unit of quantum theory.

Beginning Quantum Mechanics

Leonard Susskind and Art Friedman, *Quantum Mechanics: The Theoretical Minimum* (Basic Books, 2014). This is probably the best book to read immediately after completing the present book. The authors lead the reader into quantum mechanics using an entertaining and engaging pedagogical style.

David J. Griffiths, *Introduction to Quantum Mechanics, 2nd Edition* (Pearson, 2014). This standard undergraduate textbook requires a prior knowledge of calculus.

Richard Feynman, Robert Leighton and Matthew Sands, *The Feynman Lectures on Physics, Volume 3* (Addison Wesley [1965] 1971). This classic undergraduate text is possibly the most entertaining textbook on quantum mechanics ever written. Nobel Prize winning Feynman made major contributions to quantum theory and probably understood it better than the pioneers of the subject. While a prior knowledge of calculus is required, the book uses far more algebra than calculus.

David S. Saxon, *Elementary Quantum Mechanics* (Holden Day, 1968; Dover 2012). This book offers several exercises at the end of each chapter.

Solutions are given for selected exercises. Requires a prior knowledge of calculus.

Quantum Theory beyond Physics

Bruce Rosenblum and Fred Kuttner, *Quantum Enigma: Physics Encounters Consciousness* (Oxford, 2011).

Roger Penrose and Fred Kuttner, *Quantum Physics of Consciousness* (Cosmology Science, 2011).

Roger Penrose and Stuart Hameroff, *Consciousness and the Universe: Quantum Physics, Evolution, Brain and Mind* (Cosmology Science, 2015).

Amit Goswami, *God is Not Dead: What Quantum Physics Tells Us about Our Origins and How We should Live* (Hampton Roads, [2008] 2012).

Stephon Alexander, *The Jazz of Physics: The Secret Link Between Music and the Structure of the Universe* (Basic Books, 2016).

Diarmuid O'Murchu, *Quantum Theology: Spiritual Implications of the New Physics* (Crossroad, 2004).

John Polkinghorne, *Quantum Physics and Theology: An Unexpected Kinship* (SPCK, 2008).

Miriam Therese Winter, *Paradoxology: Spirituality in a Quantum Universe* (Orbis, 2009).

Danah Zohar, *The Quantum Self: Human Nature in Consciousness Defined by the New Physics* (William Morrow, 1991).

Leon Lederman and Christopher T. Hill, *Quantum Physics for Poets* (Prometheus, 2011).

Robert P. Crease and Alfred Scharff Goldhaber, *The Quantum Moment: How Planck, Bohr, Einstein and Heisenberg Taught us to Love Uncertainty* (Norton, 2014).

General Physics

Finally, for those who wish to learn not just quantum theory, but other areas of physics as well, there are several textbooks available. Most of these tend to be very expensive and extremely bulky, but you may find some used copies that will fit into your budget. If you do not wish to master all of physics, but simply wish to get a quick but comprehensive view of the subject, the following is an excellent book:

Art Hobson, *Physics: Concepts and Connections: 5th Edition* (Pearson, 2010).

Index

Printed in the United States
By Bookmasters